A SHORT INTRODUCTION TO

Biomedical Engineering

S N Sarbadhikari

Associate Professor of Biomedical Engineering
Amrita Vishwa Vidyapeetham
TIFAC-CORE in Biomedical Technology
Kollam 690 525, INDIA

Universities Press

Universities Press (India) Private Limited

Registered Office
3-6-747/1/A & 3-6-754/1, Himayatnagar,
Hyderabad 500 029 (A.P.), India
e-mail: info@universitiespress.com

Distributed by
Orient Blackswan Private Limited

Registered Office
3-6-752, Himayatnagar, Hyderabad 500 029 (A.P.), India

Other Offices
Bangalore / Bhopal / Bhubaneshwar / Chennai / Ernakulam
Guwahati / Hyderabad / Jaipur / Kolkata / Lucknow / Mumbai
New Delhi / Patna

First published 2006
Reprinted 2009

ISBN: 978 81 7371 558 7

Set in Times New Roman 10/12 by
OSDATA, Hyderabad 500 029.

Printed at
Printline Printers
Hyderabad 500 029

Published by
Universities Press (India) Private Limited
3-6-747/1/A & 3-6-754/1, Himayatnagar
Hyderabad 500 029 (A.P), India

Contents

Foreword

Medical care today is an integration of biomedicine, physical sciences, and technology. Every aspect of diagnosis, therapy and rehabilitation draws upon knowledge and techniques from all these areas. To present the state-of-art practices in such a multidisciplinary field requires an in-depth perception of biomedicine and engineering – a very challenging problem. Dr. Suptendra Nath Sarbadhikari, with his unique training as a medical doctor, physiologist and biomedical engineer, has in an exceptional manner addressed this issue.

A Short Introduction to Biomedical Engineering covers a very wide range of topics, but in a concise manner, on account of the special emphasis on synthesis of concepts from different areas of medicine and technology. Basics of measurements in medical practice have been dealt with, giving it a holistic approach. Therefore, the reader will have a panoramic view of the subject, as well as, details about the technologies involved. Moreover, with his background as a medical doctor, Dr. Sarbadhikari has very critically projected the optimal application of each technique and equipment, giving their advantages, limitations, and most significantly, the implications in clinical use. The presentation is in a fast-flowing and interesting style, which will serve the needs of both medical doctors and technologists desirous of getting an overview, and also of students taking courses in the field of biomedical engineering. A glimpse of the emerging future trends in biomedical engineering also evolves from the text and stimulates thinking, as well as further reading.

Sujoy K Guha, B Tech, M Tech, MS, PhD, MBBS
Chair Professor in Biomedical Engineering
Indian Institute of Technology, and
National Institute of Medical Science and Technology
Kharagpur
India.

Preface

Science is always wrong. It never solves a problem without creating ten more.

George Bernard Shaw

In reply to Alexander Pope's "A little learning is a dangerous thing", Thomas Henry Huxley asked, "…then where is the man who has so much as to be out of danger?" The philosopher Tai T'ung said, "Were I to await perfection, my work would never be published."

These great sayings might be quite dated, but they are very much (perhaps even more) valid today. Richard Feynman had warned us against the "ever expanding boundaries of ignorance", and, it is no less true for a growing field like bioengineering.

Bioengineering is a highly prospective discipline (no more an emerging one) with vital multi-disciplinary interactions. It is the application of electrical, mechanical, chemical, optical and other engineering principles to understand, modify or control biological (plant and animal – including human) systems, and also to design and manufacture products for monitoring physiological functions, assisting in diagnoses, assessing prognoses and helping in treatment of patients. In a hospital setting it is known as clinical or medical engineering.

Bioengineering is broadly divided into two components – biomedical engineering (BME) and biotechnology. The latter is perhaps more glorified due to publicised breakthroughs in genetic engineering. However, BME is no less important. Some people believe that if the 20th century belonged to the physicists, the 21st century will definitely belong to the biologists. Now, with multidisciplinary interaction becoming practicable in a country like India, branches like biophysics and BME are gaining importance and respectability.

I do not claim to bring to you the first comprehensive book on biomedical engineering. In fact, there are two wonderful books – *The Biomedical Engineering Handbook* edited by Joseph D Bronzino (CRC Press, 2000) and *The Introduction to Biomedical Engineering* (Academic Press, 1999). However, there is still a need for a shorter and more traditional textbook on BME, as a whole. So, my modest effort is to present a synopsis (bird's eye view) of the important components of biomedical engineering in a lucid way for the benefit of undergraduate and postgraduate students. Most of the topics discussed in this book are offered as a compulsory or elective paper/subject (under various names) in many streams of M Tech./ME, B Tech./BE, in Engineering (apart from courses in biomedical engineering, biosciences and engineering, medical science and technology) and Applied Physics, AMIE, Polytechnic and Physiotherapy

(BPT/MPT) courses in many institutes/universities. As some of the topics are overlapping, they will be described in detail in one part of the textbook with a cursory reference in the other portions. Questions are given at the end of each chapter. This book does not cover principles in biotechnology.

While writing the book, I found that students who have a liberal access to the Internet can actually benefit a lot by doing a "Google" search for almost any topic and download excellent animated lectures by experts in the respective fields. Further, for the student interested in recent research articles, scholar.google.com is an excellent portal. I have also referred to a few standard textbooks related to the topics, at the end of each chapter. Apart from these, the readers may visit Free Open Courseware of MIT: *http://ocw.mit.edu/index.html*. There is also a very intelligent and interesting cluster search engine: *www.vivisimo.com*

I am greatly indebted to Dr Bhagwati Prasad Haldiya (also for initiating me to Programming in C), Dr Sushmita Mitra, Dr Suman Kumar Mitra and Dr Sanghamitra Bandyopadhyay, for their liberal help in teaching me to write programs in C++.

No word of appreciation will be sufficient to express my gratitude to Prof. D M Vasudevan, presently the Principal, Amritha Institute of Medical Sciences, Kochi, and Dean, Faculty of Medicine, Amrita Vishwa Vidyapeetham, who besides encouraging me to write this book also spared his valuable time and experience to help me in any difficulty I faced while writing.

Prof. Sujoy K Guha, S R Sengupta Chair Professor, Indian Institute of Technology Kharagpur, has always been an inspiring figure to me. He had been gracious enough to guide and encourage me while writing this book and has kindly agreed to write the Foreword for this humble venture. I feel greatly privileged by this gesture.

I am also grateful to all my friends and colleagues (senior and junior) at the Sikkim Manipal Institute of Medical Sciences, Sikkim, IIT Kharagpur and Amrita Vishwa Vidyapeetham for their constant encouragement and support during the writing of this book.

The blessings of my parents (staying at Kolkata) and teachers have all made this work possible. Acknowledging the patience of my little daughter Sohini and the endurance and encouragement of my wife Dr Anindya is beyond my capacity.

Last but not the least, the editorial staff at Universities Press has been very professional in publishing this book.

However, the test of the pudding is in the eating. Therefore, I await feedback (both positive and negative), from students and teachers, for further improvement in the quality of the textbook. I humbly request the esteemed readers to send in their responses via email to supten@amrita.edu, drsupten@yahoo.com or supten@gmail.com.

Supten Nath Sarbadhikari
Amritapuri, Kollam
December 19, 2005

Section 1

Bioinstrumentation

Biomedical appliances, in its many forms, have made vast improvements in providing efficient healthcare. They may be classified into implants e.g., dental, orthopaedic, described in Section IV, Chapter12), fiberoptics (endoscopes, arterial blood/gas analysers), external aids (orthotics and prosthetics), imaging modalities (Section VI), computers systems, therapeutic/destructive artificial intelligence (AI) (Section VII), micromechanics, nanomaterials, megatronics, and so on. They can also be classified by how they function on the human body as controllers (fault-tolerant appliances like the artificial heart, the heart–lung machine or fail-safe appliances like drug infusion sets, implantable defibrillators, and nuclear radiation machines), information managers and diagnostic tools. Biosensors also fall under the category of biomedical instrumentation.

Before going into the engineering principles underlying biomedical appliances we need to learn about the electrical potentials that are present naturally within the living body. This is discussed in Chapter 1.

In 1903 Wilhelm Einthoven made the first electrical recording from the heart (ECG), initiating a tool for reproducible and reliable interpretations of the heart, and thus, ushering in the era of biomedical engineering. Some important and common biomedical instruments are discussed in Chapter 2.

Presently, biotelemetry is becoming an essential component of medical practice. It is discussed in Chapter 3.

1

Origin of Biopotentials, Electrodes, Transducers and Amplifiers

1.1 Introduction

Electrical potentials exist across the membrane of essentially all cells of the body, even in rest. However, nerve and muscle cells are noticeably 'excitable' by the generation of electrochemical impulses at their membranes during action potentials (APs). They also use these impulses to transmit along their membranes. Galvani, in the 18th century, was the first person to demonstrate that most physiological processes are associated with electrical changes.

1.2 Biopotentials

1.2.1 Resting Membrane Potential (RMP)

At rest, there exists in every membrane, the resting membrane potential, RMP (similar in magnitude to the equilibrium potential of chloride ion, E_{Cl} which is about –70 mV in nerves and about –90 mV in muscles). This potential is achieved by the lining-up of all negative ions in the inner side of the cell membrane and all positive ions on the outside, albeit the total charge on both sides of the cell membrane are equal.

Nernst (Planck) equation

The RMP is maintained by the equation

$$E_{Cl} = \frac{RT}{FZ_{Cl}} \ln \frac{[Cl_o^-]}{[Cl_i^-]}$$

where E_{Cl}^- is equal to the Nernst or equilibrium potential for chloride ion (Cl^-), *R* is the gas constant (8314.41 m J / mole K), *T* is the absolute temperature (310 K or 37°C), *F* is the faraday (96485 Coulombs / mole of charge), Z_{Cl}^- is the valence of chloride ion (–1), $[Cl_o^-]$ is the Cl^- concentration outside the cell and $[Cl_i^-]$ is the Cl^- concentration inside the cell. Converting the natural logarithm to the base 10, and replacing the constants with the numerical values, the equation becomes:

$$E_{Cl} = 61.5 \log \frac{[Cl_i^-]}{[Cl_o^-]} \text{ at } 37°C$$

The concentration ratio of chloride ions has been reversed due to multiplication by the valence (–1). Similar equations can be obtained for other ions too.

Goldman Hodgkin Katz (GHK) constant field equation

The magnitude of the membrane potential, at any instant depends upon the distribution of sodium or natrium (Na^+), potassium or kalium (K^+) and chloride (Cl^-) ions and their membrane permeability. The equation which describes this relationship fairly accurately is:

$$V = \frac{RT}{F} \ln \frac{P_{K^+}\left[K_o^+\right] + P_{Na^+}\left[Na_o^+\right] + P_{Cl^-}\left[Cl_i^-\right]}{P_{K^+}\left[K_i^+\right] + P_{Na^+}\left[Na_o^+\right] + P_{Cl^-}\left[Cl_o^-\right]}$$

where *V* is the membrane potential, *R, T* and *F* are as before, *P* denotes permeability of the respective ions, the brackets signify concentration, and subscripts i and o refer to the inside and outside of the cell respectively. As P_{Na}^+ is low relative to P_K^+ in the resting cells, Na^+ contributes little to the value of *V.* This equation also predicts that external $[Na^+]$ produces almost no change in the RMP, while an increase in external $[K^+]$ reduces the RMP.

Ion channels

K^+, Na^+, Ca^{++} and Cl^-, all have ion channels which exist in multiple forms with diverse properties. Important information about the ion channels has come extensively from studies called *patch clamp studies*.

Patch clamp studies

The technique which has allowed us to view ionic and electrochemical changes at cellular level is known as *voltage clamping.* This measure the current flow, keeping the voltage applied at a constant. Of the various types of voltage clamp available, the most popular and widely used is the *patch clamp*. Here a micropipette is placed on the membrane of a cell and forms tight seal to the membrane. The patch of the membrane under the pipette tip usually contains only a few transport proteins and they can be studied thoroughly.

Some transport proteins make up simple aqueous ion channels. Some of these are continuously open, while others are gated. The gates are opened or closed either by alteration in the membrane potential (voltage gated), or by the bonding of a specific ligand or chemical to the gate which opens or closes it (ligand gated).

1.2.2 Action Potential (AP)

The main language of the neuromuscular system is the Action Potential (AP). *AP* can be defined as the propagated, non-graded impulse, caused by changes in the conduction of ions, across the cell membrane, due to alterations in ion channels. The nerve cell membrane is polarised at rest, with the positive charges lined up along the outside of the membrane and the negative charges along the inside. During depolarisation, the membrane suddenly becomes permeable to Na^+, allowing a large number of Na^+ ions to flow into (influx) the interior of the axon, thus raising the potential rapidly and positively. The peak of the action potential is known as the *spike* or *overshoot*. Within a few fractions of a millisecond after that, the Na^+ channels begin to close and the K^+ channels slowly widen more than normal. Rapid diffusion of K^+ to the exterior (efflux) re-establishes the usually negative RMP. This is called repolarisation (of the membrane).

The Na^+ channels are opened by the positive feedback, and a vicious circle follows. The threshold potential or firing level for AP in nerves is usually about –65 mV. Shortly after an AP is initiated, the Na^+ channels become inactivated and no amount of excitatory signals there, will open the inactivation gates. During the absolute refractory period (ARP), which is about 1/2500 of a second in large myelinated nerve fibres, an AP cannot be elicited even with a strong stimulus. In the relative refractory period (RRP) following the ARP, supranormal stimuli can elicit an AP. In all normal excitable tissues, if an AP is initiated at any point of the membrane, depolarisation will travel over the entire membrane only in right conditions. They will not travel at all under adverse conditions. Also, there is no increase in amplitude with rise in strength (intensity) of the stimulus. This is the "all or none" law.

During rest (before the AP), the conductance (inverse of resistance) for K^+ is 50–100 times more than the conductance of Na^+ because of the much greater leakage of K^+ than Na^+ through leak channels. Figure 1.1 shows a typical AP. S stands for the point of stimulation, SL is the latent period and T the

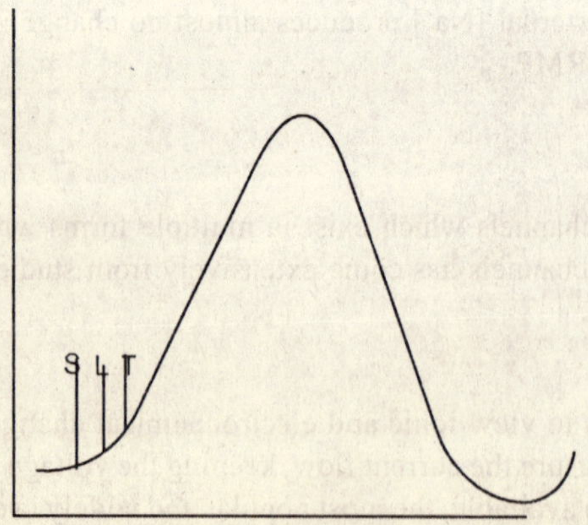

Fig.1.1 A hypothetical AP

Table 1.1 Summary of ionic changes for action potential

1. Opening of some Na^+ channels
2. Na^+ influx
3. Threshold potential or firing level is reached
4. Positive feedback open further Na^+ channels
5. Maximum Na^+ influx: Depolarisation
6. Overshoot/spike
7. Inactivation of Na^+ channels
8. Delayed opening of K^+ channels
9. Repolarisation caused by K^+ channels
10. Delayed closure of K^+ channels
11. Hyperpolarisation of membrane.

threshold potential or firing level. At the onset of an AP, Na^+ becomes activated instantaneously and allows a 500–5000 times increase in conductivity of Na^+. Then, the inactivation process closes the Na^+ channels within a few fractions of a millisecond. The onset of an AP also causes the voltage gating of K^+ channels leading them to begin opening more slowly. At the end of the AP, the repolarised RMP causes the K^+ channels to close back to normal, but after a delay, causing after-hyperpolarisation. The latent period is utilised in measuring the nerve conduction velocity (NCV) described in Section 2.4. Table 1.1 summarises the ionic changes that occur in a membrane during and after an AP.

Rheobase is the minimal electric current of infinite duration necessary to produce a twitch in a muscle. Chronaxie is the minimum interval of time necessary to electrically stimulate a muscle or nerve fibre, using twice the minimum current (rheobase) needed to elicit a threshold response. When the stimulus intensity is below the rheobase, the stimulus intensity is ineffective even at very long stimulus durations. The nerve membrane is not an ideal capacitor since it has no leakage resistance. Actually, there is a constant net outward leakage current across the cell membrane, which under resting conditions, opposes the inward flow of current due to stimulation. This is due to the outward diffusion of K+ ions via the K+ leak channels. When the magnitude of this (outward) leakage current is equal to or greater than that of the (inward) stimulating current, depolarisation can no longer occur in response to external stimulation. This occurs at the rheobase, below which, externally applied current cannot depolarise the membrane. The terms 'rheobase' and 'chronaxie' have applied importance in assessing the status of neuromuscular excitability, especially the recovery following a paralysis or injury.

Factors affecting recording of biopotentials

Many factors affect the recording of biopotentials. To cite a concrete example, in an electrogastrograph (EGG), the gastric electric signals are transformed by the electrical conductivity of the body, changing the internal gastric electrical activity (GEA) signal into a weak, soft signal very close to a sinusoidal form. The EGG wave shape also differs from a GEA internal (invasive) recording because moving the electrodes away from the gastric wall increases the gastric area from which electrical behaviour is recorded i.e., internal recordings (of migrating motility/myoelectric/motor complexes or MMC) investigate only the local GEA while an EGG gives a global view of the electrical activity in the whole stomach occurring at a given instant. The EGG wave shape is also modified by the noise introduced by electrical fields of other organs like the heart and intestine, as well as by the process of EGG acquisition (e.g., respiratory movement artifacts), filtering and amplification. Extracting quantitative information from an EGG is a technological challenge. Some of the artifacts that affect signal quality include electrode properties (active area size and construction), electrode–skin interface impedance, obesity, interelectrode distance, filtering system and gastric size.

Just as there is variation of EGG characteristics between species, there are many factors, which may affect the EGG even within a single human species. Body mass index (BMI) affects EGG stability due to the increased distance between the source of the electrical field (stomach) and the recording electrodes caused by the increased amount of fat between the stomach and the abdomen. The rate and power of the gastric electrical rhythm is not influenced by age, gender and BMI, although its stability is affected. The instability of the electrical rhythm can be calculated by the dominant frequency instability coefficient (DFIC). It has been found that this coefficient is lower in the postprandial state and in elderly women. In other words, the dominant gastric electrical frequency is higher in the postprandial state compared to the fasting state. Similarly, young people (compared to older ones) and males (compared to females) have lower dominant frequency. Again, water does not increase gastric contraction, but increases slow wave amplitude while reducing frequency. On the other hand, a solid meal increases both the frequency and amplitude of contractions. Apart from physiological conditions, pathological conditions (both gastrointestinal and non-related) also alter the EGG properties.

Visual interpretation of EGG paper recordings is imperfect and time-consuming. Digital analysis, developed recently, has improved EGG quantification. Just measuring the EGG wave amplitude or power dynamics show an increase with feeding, gastric distension and displacement. So nowadays, time–frequency measurements (running spectral analysis) are more widely accepted. This is because these measurements not only provide information about the EGG frequency and its dynamics but can also separate gastric signals from noise (especially respiration), making EGG analysis easier and more accurate.

Despite several claims of specific EGG abnormalities in clinical disorders, few have been substantiated. This is due to differences in EGG methodology and standardisation, both in the process of acquisition as also in the interpretation. However it has been shown that EGG tachygastria (increased gastric electrical activity rate) corresponds to gastric electrical uncoupling without an increase in frequency. Considerable investigation and validation of current methodology is necessary to understand the real meaning of EGG abnormalities.

1.3 Electrodes and Transducers

Electrodes transform the ionic conduction occurring in tissues to electronic conduction in measuring instruments. These may be surface or invasive (deep-seated like needle and wire or micro) electrodes. Generally, silver–silver chloride (Ag–AgCl) electrode and electrode gels are used for measuring biopotentials. Surface electrodes may have 2–10 kΩ resistance, while needle electrodes have greater resistance. Incidentally, the electrical resistance of our plasma membrane is around 7Ω. The electrodes may be used for both recording as well as stimulating biopotentials (Section 2.11).

Transducers convert various forms of energy (chemical, heat, mechanical or pressure, vibration, light, sound) into another more measurable (usually electrical) form. As most measuring devices are electrical and electronic, most physiological events are usually transformed into electronic form through transducers.

Transducers measuring displacement (e.g., goniometer) measure variable resistance and are usually passive. Strain gauge (SG) types of transducers can measure distension (e.g., of blood vessels). A linear variable differential transformer (LVDT) can measure limb movements. Velocity can be measured by potentiometers, electromagnetic (EM) and ultrasonic (US) transducers. Acceleration may be assessed by SG, EM or piezoelectric types of transducers. Flow (rheology) can be measured by hot wire, turbine, Venturi and EM transducers. Force and pressure are usually measured with SG and LVDT transducers. Pressure may also be ascertained with piezoresistors. Vibration is commonly measured with EM and piezoelectric types. Among the active or self-generating transducers, thermistors and thermocouples usually measure temperature. Humidity is normally measured with a resistance and deformation hygrometer. Illumination may be measured with a photovoltaic cell and light-dependent resistor (LDR), while chemical transducers are glass or ion-selective resistors, thick film sensors or ion-selective field effect transistors (FET). Resistors may also be of the pull-up or pull-down type. A pull-up resistor is an open collector, where one can connect a resistor of required value (depending on the amount of current one wants to draw) in series to the collector and connect the load parallel to the resistor. In the case of a pull-down resistor, one can connect a resistor of required value in series to the emitter and connect the load parallel to the resistor. Both work when the transistor is in saturation mode.

1.4 Bio (medical) Sensors

Bio (medical) sensors are used for sensing various changes in the composition of *in vivo* chemicals like ions and neurotransmitters (e.g., nitric oxide or NO), motor activity of a subject and also the cellular electrical changes. Conventional electrosensing units usually use a field effect transistor (FET), but newer methods are described later in this section.

Ortiz and Garvin [3] explain the advantage of using the high performance Nitric oxide monitoring system (developed by the Harvard Apparatus Company) for *in situ* studies. This system has a high level of specificity and sensitivity for NO, not available in conventional analysers. Some of its advantages include: (i) ease of use, (ii) high sensitivity, (iii) versatility: single or dual channel, (iv) temperature compensation assuring accuracy, (v) reliability and extended sensor life-time because of advanced tough-coat sensors, (vi) availability of several sensor designs to cover a wide range of applications including cell culture, *in vivo* measurements, (vii) availability of PC software for operation and data storage, (viii) wide pH range, (ix) high selectivity for NO, (x) non-requirement of reference electrodes as all NO sensors are combination sensors, (xi) durability as it has a mechanically-rugged membrane, (xii) selectivity and fast response time as all sensors are covered with a proprietary triplecoat gas permeable membrane, (xiii) noise-insensitivity: all sensors are completely shielded from environmental and electric noise, (xiv) adaptability: all sensors work with the 'inNO' and 'inNO-T' meters as well as a number of other currently available meters, and (xv) non-requirement of expensive chemicals as there is only one calibration procedure, simple and easy.

However, conventional sensors may be classified into thermal (calorimetric, thermometric or pyroelectric transducers) sensors for enzymatic assay, mass sensors (fluid, i.e., gas or liquid), electrochemical sensors (potentiometric, amperometric and conductimetric) and optical or optoelectronic sensors (biosensors and fluid sensors). Insoluble salt-based sensors can identify inorganic ions, while solid ion-exchange resins can detect both organic and inorganic ions and liquid ion exchanger sensors with electrostatic selection can identify both organic and inorganic molecules. Bioaffinity sensors (which sense change of local electron densities) can locate various dyes, enzymes, hormones, antibodies and receptors for hormones, as well as transport or carrier proteins. Metabolic sensors can recognise various tissues, enzymes and organelles by substrate consumption and product formation. Coupled and hybrid systems use sequences, competition, anti-interference and amplification reactions for sensing various molecules. Biomimetic sensors can make use of sound, light or stress to identify carrier-enzymes.

Commonly, sensors use biological, chemical, electromagnetic, thermal, mechanical, radioactive or other miscellaneous means for detecting changes, while the conversion phenomena utilised are physical, chemical or biological. The physical means may be further subdivided into thermoelectric, photoelectric, magnetoelectric, elastomagnetic, thermoelastic, elastoelectric, thermomagnetic, thermo-optic, or photoelastic. The chemical processes may be physical, chemical or electrochemical transformations and spectroscopic. The biological processes maybe spectroscopic, biochemical or physical transformations, or the effects on test organisms.

A subject's motor output may be measured with sensors for (i) single degree of freedom (DOF) movements like flexion–extension around the elbow and (ii) for two or more DOF movements, like coordinated hand movements, which may be either 1-D (e.g., linear transducer) or 2-D (e.g., steering wheel). Sensors for 2-D tasks, obviously, have to move with 2 DOF, as in the case of joysticks. 1-D sensors can be used for 2-D tasks by bi-manual tracking, e.g., two 1-D joysticks may be utilised for determining dynamic synergies between the two arms.

In *in vitro* neural circuits and biosensors, the electrochemical and chemoelectrical transduction properties of neurons are utilised as the basis of a cell-based biosensing unit.The synaptic connection of neuronal cells may also be used for *in vitro* neuronal circuits. Both of these require controlled neuronal cell attachment, tightly coupled to a substrate, and a sensitive substrate to monitor the changes in cellular electrical activity. A neural chip may be created with an FET-based neuron–silicon junction, as a biocomposite.

Bio-MEMS and Bio-NEMS are recent technologies in the domain of biomedical sensors. *M (micro) or N (nano) Electro Mechanical Systems* are nothing but very small electromechanical systems, which may be used for building dry (no intervening electrode paste required) electrodes with barbed micro or nano tips for fixing directly over the skin (because of the minute size, it is practically painless like a mosquito bite). Using some techniques of microfluidics, they may also be used for measuring many chemicals like serum glucose and electrolytes simultaneously on a single chip (lab on a chip). The latter is rather inexpensive due to batch fabrication and it is already available in the market. Of course, the use of Bio-MEMS is not related to just these simple sensors. They can also be used in the area of neuroprostheses for retinal or auditory implants (Section 12.3.6). This is still in the research level only. Other than their use as sensors, they are also used as actuators.

1.5 Amplifiers

Amplifiers used for biological measurements include (i) ac/dc universal for EMG, EEG, EOG, (ii) ECG or EKG with full 12 lead selection (iii) transducer/amplifier for strain gauge (e.g., BP), force, resistance, temperature devices and direct low level dc input signals and (iv) dc amplifier along with standard thermistor probes for accurate measurements of a patient's temperature.

Before proceeding further, we must make ourselves familiar with a few more terms. The *common mode rejection ratio (CMRR)* is the ratio between the amplitude of a common mode signal and the amplitude of a differential signal that would produce the same output amplitude. It can also be defined as the ratio of the differential gain over the common mode gain: CMRR = G_D/G_{CM}. Expressed in decibels, the common mode rejection is 20 $\log_{10}$ CMRR. It is a function of the frequency and source-impedance imbalance.

The *isolation mode rejection ratio (IMRR)* is the ratio between the isolation voltage, V_{ISO}, and the amplitude of the isolation signal appearing at the output of the isolation amplifier. It can also be defined as the isolation voltage divided by the output voltage V_{OUT} in the absence of a differential and common mode signal; IMRR = V_{ISO}/V_{OUT}.

Operational amplifier (op-amp) is a very high gain dc-coupled differential amplifier with single-ended output, high voltage gain, high input impedance, and low output impedance. Because of its high open-loop gain, the characteristics of an op-amp circuit depend only on its feedback network. So, the integrated circuit (IC) op-amp is a very convenient tool for the realisation of linear amplifier circuits. Time constant (tc), which will also have to be defined, is discussed in Section 2.5.

The basic requirements of an amplifier for biomedical appliances include (i) no influence on the biopotential generation or transmission, (ii) no distortion, (iii) best possible separation between signal and noise or interferences, (iv) protection of patient from electrical shock and (v) self- protection from hazards of high input voltage.

Biopotential amplifiers are crucial for biomedical measurements. The quality and information content of the signals depend mainly on the performance of these amplifiers. However, there is hardly any scope for a general purpose or universal bioamplifier, since bandwidth, sensitivity, dynamic range, CMRR and patient safety considerations call for specific amplifiers.

A common scheme for a bioamplifier will be: Source of biopotential → pre-amplifier (gain 10–50) → high-pass filter → isolation amplifier (gain 10–1000)→ low-pass filter → output

The isolation amplifier will allow for patient safety, preclude ground loops and lessen the noise.

In Section 18.2 (Fig. 18.2), we will see that the first amplifier is also known as the *pre-amplifier*. This may be a differential (balanced), carrier, chopper or phase-sensitive detector (for LVDT) type of amplifier. Pre-amplifiers usually amplify the voltage, so they require high input impedance (2–10 MW). Other specifications include isolation and protection circuitry to avoid micro or macro shocks with a limited bandwidth to give high signal-to-noise ratio (SNR).

Since biopotentials are of rather low amplitude, the gain of bioamplifiers must be high (1000 or more). The CMRR of bioamplifiers ought to be substantial enough to alleviate common mode signals, as the common mode voltage is quite high compared to the biopotential which needs to be measured by the bipolar differential electrodes. Often unprocessed amplitude magnitude is taken as a result for biopotentials (like EKG). A calibration standard is, therefore, essential. For oscilloscopes (wideband applications), chopper stabilised dc amplifiers are used. It has three amplifiers in the module, one ac amplifier for signals higher than 20 Hz, one dc chopper for signals from 20 Hz down to dc and a wideband feedback stabilised dc amplifier.

Table 1.2 Some characteristics of biomedical signals.

		Usual recording parameters						Signal characteristics		
Name	Origin	Cut off frequency (Hz)		Sensit -ivity	Time constant	Paper speed	50 Hz filter	Freq- uency	Amp- litude	Voltage range
		Low	High	μV/cm	(s)			(Hz)	(μV)	
ECG	Heart	1	75	1000	0.01	25 mm/s	In	0.05–500	1000	10μV – 5mV
EEG	Brain	1	75	100	0.10	2.5–5 mm/s	In	0.1 – 100	40 – 60	2 – 200mV
EMG	Muscle	3	75	50	0.04	25 mm/s	Out	5 – 200	100 – 1000	25μV – 5 mV
EOG	Eye	0.3	35	200	0.24	2.5–5 mm/s	In	dc – 100	500	10μV – 3.5 mV

The pre-amplifier is connected to a second (main) amplifier, which leads to the driving stage. The gain and frequency response of the main amplifier is adjusted to get the proper stylus deflection and frequency response.

Table 1.2 shows some characteristics of biomedical signals.

In the next chapter we shall deal with specific biopotentials and their measurements.

1.6 Further Reading

[1] Raja Rao, C and Guha, S K. *Principles of Medical Electronics and Biomedical Instrumentation.* Universities Press (India) Ltd. 2001.

[2] Khandpur, R S. *A Handbook of Biomedical Instrumentation.* 2nd edition. Tata McGraw Hill. 2003.

[3] Ortiz, P A and Garvin, J L. Interaction of O_2 and NO in thick ascending limb. *Hypertension* 39 [part2]: 591–596. 2002

[4] Cromwell, L, Weibell, F and Pfeiffer, E A. *Biomedical Instrumentation and Measurements.* Prentice Hall of India Ltd. 1993.

[5] Aston, R. *Principles of Biomedical Instrumentation and Measurements.* USA: Merrill Publishing Co. 1990

1.7 Questions

1 Descriptive questions

1. What is the ionic basis of the resting membrane potential (RMP) and the action potential (AP)?
2. What are the advantages and disadvantages of recording internal biopotentials from the surface of the body?
3. What are the different types of transducers? Give examples of each type and describe their function in detail.
4. What are the basic requirements for an electrode system to record biopotentials?
5. What are the various types of bioamplifiers?

2 Write short notes on

1. 'All or none' law
2. Absolute refractory period
3. Relative refractory period
4. Chronaxie
5. Rheobase
6. Force transducers
7. Surface electrodes
8. Bio-MEMS

3 Multiple choice questions

Choose the single best response.

1. Electrical resistance of plasma membrane is

   (A) 7Ω (B) 10Ω (C) 100Ω (D) 1000Ω.

2. Resting membrane potential (RMP) is close to the equilibrium potential of

   (A) Na^+ (B) Mg^{++} (C) Cl (D) Ca^{++}

3. Action potentials are produced by

   (A) Na^+ influx (B) K^+ efflux (C) Na^+ and K^+ influx (D) K^+ influx

4. Transducers can convert the energy form of

   (A) electrical to chemical (B) mechanical to electrical
   (C) chemical to mechanical (D) all of the above

5. Bio (medical) sensors can sense the energy form of

   (A) electrical (B) physical (C) chemical (D) all of the above

Answers

1. A; 2. C; 3. A; 4. D; 5. D

2

Some Important Biomedical Equipment

2.1 Introduction

In 1903 Wilhelm Einthoven made the first electrical recording from the heart (ECG), initiating a tool for reproducible and reliable interpretations of the heart, and thus, ushering in the era of biomedical engineering. Since then, innumerable appliances have been invented to bring about more efficient health care ranging from dental implants, to the sophisticated heart–lung machine to the recent use of nanomaterials and AI (artificial intelligence) . In this chapter we will discuss certain important and common biomedical instruments used routinely by hospitals and health practitioners.

2.2 Electrocardiography (ECG or EKG)

The heart comprises four chambers (two atria—left and right, designated as LA and RA respectively, and similarly, two ventricles denoted by LV and RV). It has an inherent rhythmicity, which originates from the sino-atrial (SA) node which is the natural cardiac pacemaker. During a diastole (relaxation), the heart fills with blood and during a systole, (contraction) the heart empties its chambers. The right side of the heart contains deoxygenated (impure) blood (coming from the veins of the body into the RA, passing through the LA and flowing to the lungs for oxygenation), while the left side of the heart contains oxygenated blood (coming from the lungs into the LA and passing through the LV into the systemic circulation). The complete series of events taking place in the heart, starting from any point (e.g., atrial systole) up to that successive point is called a cardiac cycle.

Electrocardiography (ECG or EKG) is the recording taken from the body surface of the electrical changes that occur within the heart during a cardiac cycle. The instrument used to take this recording is called the *electrocardiograph* and the tracing (on paper) is called the *electrocardiogram*.

The recording electrodes (or leads) are bipolar (I, II, III—limb leads) or unipolar (aVR, aVL, aVF—augmented or Goldberger, and chest $V_1 - V_6$). These are called the standard 12 leads. There are other unconventional leads like $V_7 - V_{12}$ and esophageal leads for specific purposes.

For setting limb leads, the right leg (RL) is taken as the ground, the left leg (LL) is always positive and the right arm (RA) is always negative.

The leads are designated as: I (LA-RA), II (LL-RA) and III (LL-LA), aVR (RA), aVL (LA), aVF (LF) V_1 (right 4th intercostal space (ICS) at the sternal edge), V_2 (left 4th ICS at the sternal edge), V_3 (mid-way between V_2 and V_4), V_4 (mid-clavicular line in the left 5th ICS), V_5 (anterior axillary line horizontal to V_4), V_6 (mid-axillary line horizontal to V_4)

The paper strip usually moves at a speed of 25 mm/s i.e., each mm of strip (*x*-axis) denotes a 0.04-second duration and 10 mm on the *y*-axis denotes 1 mV of current. A positive (upwards) deflection implies that the current is moving towards the lead and a negative (downwards) deflection implies that the current is moving away from the lead. Figure 2.1 shows a typical EKG tracing with P, R, Q, S and T denoting the waves formed.

A P wave is caused by atrial depolarisation, a QRS complex is due to ventricular depolarisation and a T wave is due to ventricular repolarisation. An atrial repolarisation is usually superimposed by a QRS complex.

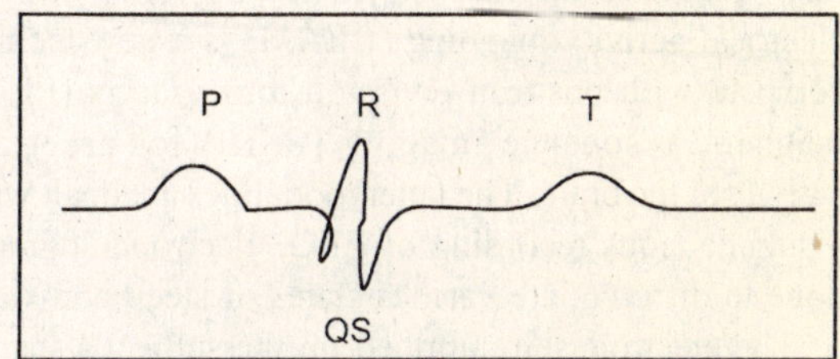

Fig. 2.1 A typical EKG waveform

In an ECG, we must first determine the heart rate, rhythm and axis. We will then get an idea of what more to look for. The ECG paper has 'small squares' at intervals of 1 mm and 'big squares' at intervals of 5 mm.

For determining the heart rate from an ECG, we must measure the interval between two successive R waves (R–R). If it is measured in seconds, 60/ R–R will give the heart rate in beats per minute (bpm). Otherwise, the number of big squares between two successive R waves in the graph must be divided by 300 (or 1500, if we are using the small squares). Once we know the heart rate, we can determine whether the subject has *tachycardia* (heart rate >100 bpm) or *bradycardia* (heart rate < 60 bpm). A very fast heart rate may lead to flutter and fibrillation, the latter being a potentially lethal irregularly irregular cardiac rhythm. Flutter or fibrillation may occur either in the atria or in the ventricles. Defibrillation is described in Section 2.7.

The PR interval is measured from the beginning of the P wave to the beginning of the QRS complex. The P–R interval is normally 0.12–0.20 second and denotes the conductivity of impulses in the heart. The normal QRS complex is usually 0.06–0.11 second. The S–T segment (elevation or depression), the Q–T interval and T waves are also to be noted.

As an ECG purveys a record of both atrial excitation (P) and ventricular excitation (QRS), it is useful for analysing cardiac dysrhythmias. P waves are generally best seen in lead II or in the right-sided chest leads (V_1), and these leads should be particularly checked for a longer duration (long lead II) of time.

Apart from arrhythmias, an ECG can be helpful in detection and localisation of changes in the myocardium due to coronary artery or ischemic heart disease (especially serially), detection of hypertrophy of the walls of the atria and ventricles, detection of pericardial disease and also detection of electrical changes in the heart following general metabolic changes.

A 24-hour Holter monitoring (portable) of ECG, an exercise-stressed (e.g., treadmill test or TMT) ECG are all variants of the conventional ECG which provide much more information than the standard ECG. Using a cathode ray oscilloscope (CRO) and a computer interface, the ECG data can be digitized, analysed, as well as transmitted electronically for telemedicine (Section 3.3) purposes.

2.3 Electroencephalography (EEG)

Electroencephalography (EEG) is the recording from the scalp of the electrical changes that occur within the brain, in sleep and activity. The instrument used for the recording is called the *electroencephalograph* and the tracing (on paper) is called the *electroencephalogram*. As the electrical activity of the brain is very less (less than 100 mV range), it has to be amplified and seen on paper or digitally. EEG paper recordings, unlike ECG recordings, are not very objective. Therefore, a qEEG (quantitative EEG) is now being used in many places to increase the EEG's objectivity and reliability. Brain electrical activity mapping (BEAM) is a new technique, which may compete with positron emission tomography (PET) and functional magnetic resonance imaging (fMRI) for ascertaining functional activity of the brain. The latter modalities are dealt with in Section 18.5. Using multi-channel polygraphy, a simultaneous recording of EEG, electrooculography (EOG) and electromyography (EMG) may be done to differentiate various states of sleep and wakefulness.

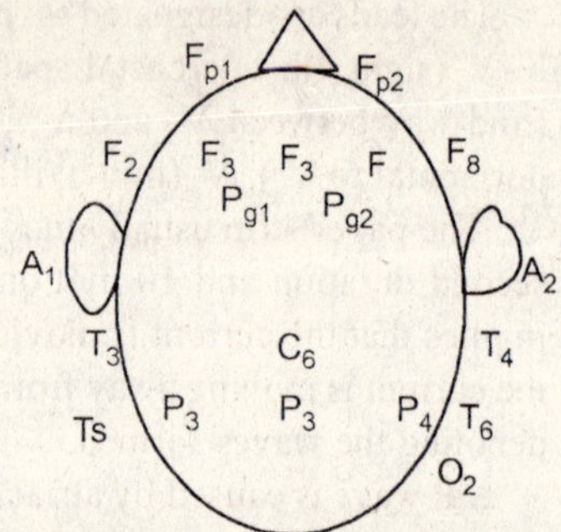

Fig. 2.2 EEG Leads

The electrodes are applied on the subject's scalp as a 10–20 system for 21 electrodes (maximum 32 channels). The 10–20 system is based on the relationship between the location of an electrode and the underlying area of çerebral cortex. Each site has a letter (to identify the lobe) and a number or another letter to identify the hemisphere location. The letters F, T, C, P, and O stand for frontal, temporal, central, parietal and occipital. Albeit there is no 'central lobe', this is just used for identification purposes. Even numbers (2,4,6,8) refer to the right hemisphere and odd numbers (1,3,5,7) refer to the left hemisphere. The z refers to an electrode placed on the midline. The smaller the number, the closer the position is to the midline. The '10' and '20' refer to the 10% or 20% interelectrode distance. EEG electrodes are placed on the scalp at 10 and 20 per cent of certain measured distances. If a measurement is made around an adult's skull, a circumference distance would be determined to be around 55 cm for an average-sized person. Therefore, 10% of this measurement or 5.5 cm would be used to determine precise locations around the skull. An anatomic landmark is needed to know where to start with the measurements. The landmarks are the ear canals, the bridge of the nose (nasion) and the little bump at the very back of one's skull (inion). The skull may be different from patient to patient. A pediatric or adolescent may be smaller than an adult, and different adults have slightly different size heads (and brains). The percentage relationship remains the same for the location of the internal brain lobes. The 10-20 system is based generally on the relationship between the location of an electrode and the underlying area of cerebral cortex, although exact relationships can only be determined with MRI confirmation of the exact locations of the various parts of the brain.

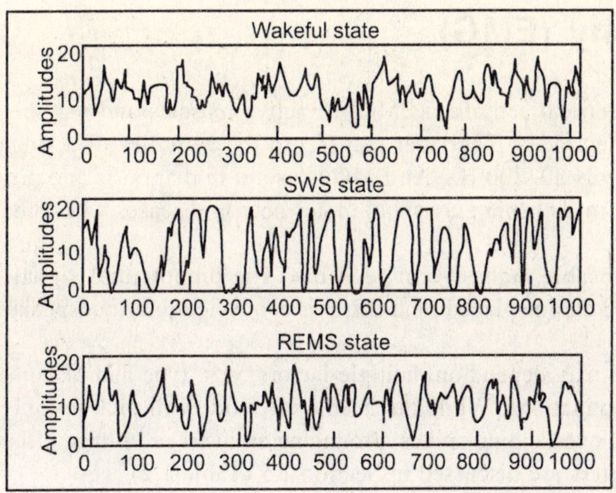

Fig. 2.3 EEG waves

According to the position (Fig.2.2), they are denoted as Fp (fronto-parietal or pre-frontal), F (frontal), C (central), T (temporal) and A (auditory). Conventionally, the odd numbers (1, 3, 5, 7) are placed on the left half of the brain (left hemisphere) and the even numbers (2, 4, 6, 8) are placed on the right half of the brain (right hemisphere). The 10–20 system is inadequate for the modern 256-channel EEG machines.

EEG waves (Fig.2.3) are classified in accordance with their frequency.

Delta (δ) waves These are of frequency less than 4 Hz (Hertz or cycles per second) and occur when the subject is in deep sleep.

Theta (θ) waves These are of frequency 4–7 Hz and usually not found in adult human beings. Interestingly, during hypnosis, it may be the predominant waveform.

Alpha (α) waves These are also known as **Berger rhythms** in honour of the inventor Hans Berger. They are of frequency 8–12 Hz and usually found in the awake but quiet (peaceful) state, like when the subject's eyes are closed or while meditating.

Beta (β) waves These are waves of frequency 13 Hz and above, found in most of our wakeful, alert and active periods and also during rapid eye movement (REM) sleep. Commonly, waves of frequency 13–18 Hz are known as β_1 wave, while waves of frequency up to 30 Hz are known as β_2 waves.

Gamma (γ) waves These are waves with a frequency range of 40–90 Hz (some people prefer to increase the range to 30–200 Hz) and are now taken to be important for cognitive functions. A 40 Hz oscillation is also known as a 'binding oscillation' as it is supposed to be most important for integrating our multimodal sensory inputs into a coherent perception.

The EEG is of particular value in the investigation of epilepsy, dementias, localisation of brain tumours and other intracranial space occupying lesions (ICSOL). It is also used in sleep research.

Brain maps or BEAM are discussed in Section 6.4.1.

2.4 Electromyography (EMG)

An electromyograph records the electrical activities (EMG) of active muscles and may be recorded with surface electrodes or needle electrodes. The latter can record the activities of a single muscle fibre. The reference range of activity is 20–200 Hz. An EMG is useful in diagnosis and prognosis of peripheral nerve injuries, myotonia, myasthenia gravis and motor neuron diseases where the rate and rhythm of nerve impulses vary.

Surface electrodes may be disposable, adhesive or reusable. The biopotential signal from the contracting muscle may be amplified and displayed on a CRO or connected to a loudspeaker through an audio-amplifier.

The amplitude of an EMG signal may depend on multiple factors, viz., type and placement of the electrodes and degree of muscle contractions. While the needle electrode will pick up only singular spikes, surface electrodes will catch overlapping spikes, producing an average voltage effect.

Some of the comparative parameters are discussed in Section 1.5 (Table 1.2).

Pre-amplifiers used for EMG are usually of moderate bandwidth and parametric in nature. The input impedance of the pre-amplifier is commonly greater than 100 mW. A narrow passband filter (Section 18.3) is usually preferred for recording an EMG.

The standard EMG has two cathode ray tubes (CRT) for display and recording. Applying beam switching, one to four traces may be obtained, as desired. Since EMG signals persist for more than 200 ms, variable persistence is helpful for visualising EMG activity.

To quantify the muscle activity (tension), it is often helpful to integrate the EMG signal. Therefore, an integrator is incorporated for converting all negative deflections (moving away from the electrodes) to positive ones.

Another accessory for an electromyograph is a muscle stimulator, which purvey single, dual or a train of pulses for external triggering. This triggering effect can be modified by changing the stimulus amplitude, duration, repetition or delay. The output too may have either a constant voltage or a constant current. The latter is usually fixed between 0 and 100 mA. The former may have square wave pulses ranging from 0–500 V in amplitude, 0.1–3.0 ms in duration and having a frequency up to 100 Hz.

As with other analog bioelectric signals (barring EKG) quantification is very much necessary, especially for detecting early changes in the EMG parameters pointing towards initiation and/or progress of a neuromuscular pathology and also during recovery from such pathologies.

Nerve conduction velocity (NCV) is measured with an EMG machine and a nerve stimulator (which may be in-built). The velocity of nerve conduction can be measured by dividing the distance between the stimulating and recording electrodes and the latent period. For example, if the distance between the two electrodes is 6 cm and the latent period is 2 ms, the conduction velocity is 6 cm / 2 ms or 30 m/s.

2.5 Polygraphy

While taking a polygraph, for example, polysomnography or sleep recording (ideally it should have been called polyhypnography or multisomnography to avoid the Greek–Latin hybrid word), many parameters like pulse, blood pressure (BP), respiratory rate, sweating (galvanic skin response, GSR or skin resistance response, SRR or psycho-galvanic response, PGR), ECG, heart rate, plethysmography (regional blood volume), temperature, EEG, EMG, electrooculogram (EOG) can be recorded simultaneously in the various channels. A polygraph is otherwise known as a 'lie detector' as it is sometimes used to detect if a subject (e.g., suspected criminal perpetrator) is lying. The principle is that fright (while telling lies) causes sweating, which will reduce the SRR and increase the heart rate and BP. However, the judicial system does not accept the polygraph as foolproof evidence of guilt.

In this instrument, the amplifiers and transducers in the separate channels may be different in accordance with the need. There will also be a channel for a time-marker (placing a tick mark at a preset fixed time interval). Tuning forks with a frequency of 100 Hz have been used for time-marking.

While recording signals, we come across the term *time constant,* which refers to the time taken for a signal to decay exponentially (i.e., to $1/e$ or 1/2.718 of the maximum or the peak of the signal). The time constant (tc) is usually inversely related to the sensitivity of the recording (biopotential) coupler.

The relative paper speeds and sensitivities for some common signals are presented in Table 1.2, Chapter 1. For some other couplers they may be of the following range:

Pulse coupler (10, 5000)*
Respiration coupler (2.5, 200)
Strain gauge coupler for muscle contraction (5, 100)
" for arterial BP (5, 50)
" for volume (2.5, 50)
" for temperature (0.25, 100)
Isotonic muscle contraction coupler (5, 500).

* Paper speed is in units of mm/s, followed by the sensitivity, measured usually in μV/cm

2.6 Magnetography

It has been found that recordings of the magnetic field of an electromagnetic radiation are more reliable than recordings of the electric field. *Magnetoencephalography (MEG)* and *Magnetocardiography (MCG)* are the recordings of the magnetic field from the electromagnetic radiations of the brain and heart respectively. These are more reliable as the magnetic waves are not attenuated by body fluids and intervening tissues, though they are cost-prohibitive. Further, the axes of the magnetic fields are different from those of the electrical fields, thus more information can be obtained by combining the modalities like EEG/MEG and ECG/MCG. See Chapters 16 and 19 for a comparison between MEG / PET / fMRI. Table 2.1 gives a comparison between an MEG and an EEG.

The common rhythms found in the wakeful state in an MEG are:

(a) the parieto-occipital a rhythm with α peak around 10 Hz, recorded above the parieto-ocipital sulcus (POS);
(b) the μ rhythm which dampens during movement (as in EEG) with suppression starting 1–2 seconds before movement—bilateral suppression (contra-laterally dominant) being common. The two main frequency components are around 10 Hz and 20 Hz and
(c) the τ (tau) rhythm of the temporal lobe with a 8–10 Hz oscillatory activity best seen in a planar gradiometer recording just over the auditory cortex.

On rare occasions, some other rhythms are also found like the θ (5–7 Hz) rhythm and the stimulus-related 40 Hz activity from the auditory-evoked field by thalamocortical stimulation. During sleep, there is usually a magnetic spindle (11–15 Hz, 0.3–2 seconds duration, during stage 2 sleep). A typical magnetic K complex lasts for 0.8–1.2 seconds and during stage 3 & 4 sleep (slow wave sleep or deep sleep), 0.5–2 Hz polymorphic MEG waves are found.

Table 2.1 Comparative advantages and disadvantages of MEG & EEG

Parameters	MEG	EEG
Availability	rare	common
Primary current (EM fields)	same	same
Attenuation	minimal	great
Effect of reference electrode	nil	dependent
Dipolar EM fields	rotated 90° compared to each other	rotated 90° compared to each other
Isocontour lines	more tight	less tight
Model for interpretation	realistically shaped brain model will suffice	multi-compartment model with known conductivities and shapes for brain, skull, csf and scalp required
Selectivity to tangential currents in the presence of several simultaneous sources	possible and important practical advantage for SSEPs	not so
Interpretation	easier	best interpretation by combining with MEG
Cost	prohibitive	cheap

SSEP = somato-sensory evoked potentials; csf = cerbro-spinal fluid.

2.7 Defibrillator

The causes of ventricular fibrillation (VF) include myocardial infarction, dyselectrolytemia (change in normal ionic composition of blood) and electrocution. As mentioned in Section 2.2, VF is potentially lethal. The irregularly irregular rhythm, therefore, has to be reverted to a more synchronous rhythm. This is done by applying a high energy shock to the heart, through a defibrillator.

The electric shock is applied either externally (closed chest or trans-thoracic) or internally (open chest).

Commonly, the trans-thoracic (dc) defibrillators are nothing but energy storage capacitors, charged slowly from ac mains or from a battery, and discharged very rapidly (ms), through a silicon controlled rectifier (SCR) in series, during the process of defibrillation. Earlier ac defibrillators have been tried, albeit the drawbacks include repetitive attempts in VF, and inefficiency for correcting atrial fibrillation.

Usually rectangular or trapezoid waveforms are designed in clinical defibrillators. Successful defibrillation has a better correlation with the magnitude of stored energy in the capacitor, rather than with the voltage. So, a dc defibrillator is calibrated in watt-seconds (Ws) or joules (J) denoting the stored electrical energy. Commonly, a defibrillation may purvey output varying up to 400 Ws, albeit 350–360 Ws is the standard for external defibrillation, while 100Ws is sufficient for internal.

$$E(\text{Ws}) = \frac{1}{2}CV^2$$

where energy (Ws) is related to capacitance C (Farads) and voltage V (Volt), such that, if a 16 mF capacitor is used, for a full discharge of 400 Ws, the capacitor must be charged to 7000 V.

The chest wall impedance varies with the size and shape of the electrodes, paddles, energy of discharge, number and frequency of previous counter-shocks, and also the interface materials between the electrode and the chest wall. The electrodes may be placed anterior–anterior or anterior–posterior.

A synchroniser is essential for cardioversion, where defibrillation is synchronised with the cardiac cycle, in a safe period. It is used in the case of atrial fibrillation or ventricular or supraventricular tachycardia. Here, defibrillation is done before the T wave appears (20–30 ms after the R wave). The EKG amplifier triggers a time delay circuit to produce the shock at the right moment, so that correction of atrial fibrillation does not lead to VF or other abnormal rhythms.

Automatic implantable defibrillators, though theoretically possible, are still not used for routine clinical purposes.

2.8 Dialyser

The kidneys clean our body of harmful waste products like urea, creatinine and ammonia, and also maintain the water, electrolyte and acid–base balance of the body (besides secretion of hormones like rennin, erythropoietin and Vitamin D).

When the kidneys become non-functional due to some disease, its functions have to be done artificially through a process called *dialysis*. Dialysis may be peritoneal or hemodialysis.

Peritoneal dialysis may be continuous and ambulatory (CAPD), thus making it possible to perform the dialysis at homes. Here the peritoneum (a double-layered sac covering most of the abdominal viscera in our body) acts as the semi-permeable membrane necessary for filtering the waste products of metabolism. However, it is a rather costly appliance.

Generally, for long-term (up to ten years or more) dialysis, a permanent intra-peritoneal catheter is inserted. It is useful in the case of young children with renal failure, geriatric patients with cardiovascular instability and diabetics. In this method, two litres of sterile isotonic dialysate are introduced through the catheter and kept there for about six hours. It is then drained and fresh dialysate pumped in, four times a day.

Hemodialysers may be of the coil, parallel plate or hollow fibre type. The concentration gradient across the membrane determines the amount of waste products removed. To achieve the maximum gradient, the concentration of waste products in the dialysate must be maintained at zero. This may be obtained by using the dialysate only once or by applying the counter-current exchanger multiplier concept of the biological kidney and making the dialysate enter the kidney where the waste concentration of blood is lowest.

Ideally the membrane for dialysis (usually made up of cuprophan or cuprammonium regenerated cellulose) must be highly permeable to water, organic metabolites and ions, while retaining large negatively charged molecules like plasma proteins. It must have sufficient wet strength to resist tearing, and needless to say, like all biomaterials, be inert and non-toxic. To be widely accepted, it must also be cost effective.

The hemodialysis machine produces a warm dialysate that is circulated through the external dialyser assembly. The cycling of blood from the patient and back, through the artificial kidney or dialyser is also controlled by it. So, its main functions are:

(a) mixing the dialysate
(b) monitoring the dialysate
(c) pumping the blood mixed with anti-coagulants
(d) monitoring the presence of air or fat in the blood (to avoid embolism) and the drip chamber pressure
(e) monitoring the ultrafiltration rate

The essential parts of the machine comprise proportioning pumps (usually fixing the water–dialysate ratio at 35:1), a dialysate thermostat (maintaining the temperature between 36 and 40°C), a conductivity measuring cell, an effluent pump (a fixed flow, motor driven gear pump, creating negative pressure upon the dialysate), a bubble trap (for measuring venous pressure), a blood leak detector, a flow meter, a blood pump, and a heparin (anti-coagulant) pump.

2.9 Diathermy

The origin of the word 'diathermy' is Greek for 'through' (*dia*) and 'heating' (*thermy*). Diathermy may be used for surgery, as well as for physical therapy (Section 2.11). While in the former, the current used may be of lower frequency (higher wavelength) 1–3 Hz, in the case of the latter, a short wave current of about 27.1 MHz frequency is used and is termed *short wave diathermy.* Sometimes microwave and ultrasound waves are also used for physical therapy diathermy.

We shall discuss surgical diathermy in this section. The common surgical uses of electrical heating include *electrotomy* or cutting, *fulguration* or superficial tissue ablation without affecting deeper structures, *dessication* or drying up and coagulation of tissues.

The effects are achieved by the passage of high frequency current through the sharp edge of the wire/band loop/needlepoint into the tissue. The current concentrates at the point, tearing the cells below by boiling the intra-cellular fluid. Meanwhile, the indifferent electrode establishes a large contact area with the patient's body surface, dissipating the heat. In this way electrosurgical separation of the tissues occurs.

The advantages include its effortless and instantaneous action of tissue separation and coagulation of small bleeders.

The chief component includes a high frequency power oscillator, usually generating 1.75 MHz current. At the operation site, usually a small cross-section (less than 1 mm), a high temperature (may be up to 1000 °C) arc is applied to achieve a high current density.

In the common bipolar type, the high potential terminal of the machine is connected to the cutting electrode that is mounted on an insulated handle. The cutting electrodes vary with the applications and may be a needle, an angulated lancet, a straight lancet, a wire loop or an angulated band loop types. The switch for operating the machine is usually at the finger tip of the handle or at the foot end. The low potential terminal is connected to the indifferent electrode, comprising a lead plate wrapped in a cloth bag soaked in saline solution and strapped behind or on the thigh of the patient. Alternatively, a flexible naked stainless steel plate may be placed on the operating table, under and touching the patient.

The possible adverse effects of diathermy include burns, electrocution and explosion due to faulty equipment or connections.

2.10 Lasers in Medicine

Lasers (Light Amplification by Stimulated Emission of Radiation) are devices which produce coherent light with a narrow range of wavelengths. They are now quite commonly used, both for diagnostic and therapeutic purposes. Table 2.2 compares and contrasts some common types of lasers used in medicine and surgery.

They may be used for vapourising, cutting or coagulating body tissues. They have the advantage of being well-directed and focussed. They light and heat up quite rapidly, although the intensity may be rather high. But they turn off equally fast. This is a great advantage to surgeons. Further, there is no contact between the source and the sample irradiated. The absorption of heat and penetration to the far reaches of the body depend on the tissue properties.

The importance of lasers has been recognised by the scientific community in India. All scientific departments of the Government of India like the Department of Atomic Energy (DAE), the Department of Electronics (DOE), the Council of Scientific and Industrial Research (CSIR), the Department of Science and Technology (DST), the BRNS (Board for Research in Nuclear Sciences), the ILA (Indian Laser Association), have major programs in this area.

In 1991, the Science Advisory Committee to the Prime Minister prepared a status report on lasers in India and made recommendations for a National Laser Program (NLP), which was started by the Department of Atomic Energy in the VIII five-year plan. The programme aims at promoting 'Lasers and their applications' by funding major and small projects in academic institutions, hospitals and national laboratories all over India.

The National Laser Program has the following on-going projects.

(a) Laser R & D programs at the Centre for Advanced Technology, Indore.
(b) Laser R & D programs at the Solid State Physics Laboratory, New Delhi.
(c) Production Unit at CAT for manufacture of lasers and related equipments.
(d) Laser R & D at academic institutions and other national laboratories.

These projects have been divided into five groups namely: laser materials, laser physics and technology, industrial applications, medical applications and manpower training in laser physics and technology.

Table 2.2 Comparison of some common types of lasers in medicine

Parameters	Nd–YAG	CO_2	Dye	Ruby	Ar (Al_2O_3– 0.05% Cr)	He–Ne	Kr
Phase	solid	gas	solid	solid	gas	gas	gas
Wavelength, nm	1064 (infrared or ir)	10640 (ir)	400–700 (visible or v)	693 (v)	490–520 (v)	630–3390 (v)	350 (ultraviolet or uv)
Peak wavelength (optical absorption, nm)	554, 970	> 2.94	431	220–1000	≈500	several	≈ 350

Table 2.2 Continued

Parameters	Nd–YAG	CO_2	Dye	Ruby	Ar	He–Ne	Kr
				$(Al_2O_3$–0.05% Cr)			
Power (W):							
CW	50	200	5	5	5	0.1	5
Peak	1000	75000	50	50	100	2	100
Type of beam	Q-switched continuous	Continuous and pulsed	Pulsed	Pulsed tunable, pulsed	Continuous,	Continuous	Continuous
Relative strength of optical absorption	++ / +	++++	+++	++++	++	++	++
Lasing species	Nd3t in YAG lattice	CO_2 molecule	Hole-electron pairs at diode junction	Chromium-Al_2O_3	Argon	Helium Neon	Krypton
Uses	Soft tissue ablation e.g., in ophthalmo-logy, neuro-logy and dermatology	Soft tissue ablation e.g., in microsur-gery, neuro-surgery and obstetrics and gynecology	Soft tissue ablation	Soft tissue ablation e.g., ophthal-mology	Neuro & general surgery (photo coagulation)	Diagnostics	Ophthalmo-logy

2.11 Physical Therapy Equipment

Short wave diathermy (surgical diathermy has been dealt with in Section 2.9), having a wavelength of 11m, is used for its effects on pain reduction. The high frequency ac does not stimulate nerves or muscles enough to cause contraction, but can relieve pain by causing vasodilation and removing pain producing substances like prostaglandins and kinins.

Microwave diathermy uses 12.25 cm wavelength and 2450 MHz microwaves. Here, unlike the pads used in short wave diathermy, an emitter is directed towards the body portion under treatment. Therefore, no tuning is necessary.

Ultrasonic (US) diathermy actually acts as a micromassage by causing high frequency vibration. Usually it is used to alleviate pain in peripheral neuritis, arthritis and related disorders like frozen shoulder, and skin ulcers. Usually, an electrical power of less than 3 W/cm^2 is needed. The controlling parameters include US frequency, intensity and duration of the exposure. Both constant and pulsed modes are used for acoustic oscillations.

The other common physical therapy methods for treating with heat include wax bath (with warm liquid paraffin), UV (uvr) and infrared radiation (irr). However, these methods may cause discomfort and/or skin burns in the patients, before adequate heat has reached the deeper affected tissues.

Apart from the diagnostic uses of electrical apparatus (e.g., NCV described in Section 2.4), stimulating electrodes may be used for iontophoresis, i.e., a galvanic current may introduce ions (e.g., of drugs) through the skin by electrolysis.

For pain relief, two methods are used, more or less on the similar principle known as '*Gate Control Theory of Pain*'. This theory proposes a 'gate' in the dorsal horn of the spinal cord, through which pain impulses reach the lateral spinothalamic tracts (LSTT), which, in turn, carry the sensation to higher centres of perception. If simultaneously large fibre afferent neurons (type II or Ab), carrying the touch sensation, are stimulated, the pain fibres (type IV or C) are inhibited (thereby reducing the sensation of pain). Collateral fibres from the dorsal column touch fibres enter the Substantia Gelatinosa and their outputs inhibit the collaterals or interneurons presynaptically. This theory is utilised in the transcutaneous electrical nerve stimulator (TENS) and electroacupuncture, a renovation of the traditional Chinese system of medicine. In the case of TENS, either a square or a spike wave up to 76 mA and 12 to 100 pulses (of about 0.5 ms) per second is applied. However, other systems may have other specifications like 0.1 to 1.0 ms pulse between 20 and 200 Hz, amplitude up to 120 V and maximum current output of 25 mA. Skin surface application of electrical stimulus is achieved by carbonised rubber electrodes, applied with a tincture of Benzoin interface for reducing irritation.

Traction is a mechanical method for relieving load/pressure and pain (e.g., in spondylosis). It is often used for correcting orthopedic deformities or setting of a particular treatment for correcting bone fractures.

2.12 Further Reading

[1] Gevins, A.S. Quantitative human neurophysiology In *Techniques in Human Neurophysiology*. Ed. H.J. Hannay. New York: Oxford University Press. 1986. 419–456.

[2] Jansen, B.H and Brandt, M.E, Eds. *Nonlinear Dynamical Analysis of the EEG*. Singapore: World Scientific. 1993.

[3] Raja Rao, C and Guha, S.K. *Principles of Medical Electronics and Biomedical Instrumentation*. India: Universities Press Pvt. Ltd. 2001.

[4] Melzack, R.The tragedy of needless pain. *Sci. Am*. 262: 27. 1990

[5] Weiner, S.L. *Differential Diagnosis of Acute Pain by Body Region*. McGraw Hill. 1993.

2.13 Questions

1 Descriptive questions

1. What are the 12 standard leads for recording ECG? What do P, Q, R, S and T denote and what is the reference ranges of the various features (P, QRS, PR, QT and RR) of EKG?
2. What are the difficulties in interpreting EEG?
3. What is the advantage of polygraphy over other recording devices?
4. How are magnetic fields used for judging brain activity?
5. What are the various methods available for treating end stage renal (kidney) disorders ?

2 Write short notes on

1. Cardiac defibrillator
2. Isoelectric EEG
3. Alpha rhythm

4. EMG
5. Surgical diathermy
6. MEG
7. MCG
8. Fusion of ECG/MCG
9. Lasers in medicine
10. Pain relieving instruments

3 Multiple choice questions

Choose the single best response.

1. The frequency of β waves in EEG is
   (A) < 4 Hz (B) 4–7 Hz (C) 8–12 Hz (D) None of the above.

2. Berger waves (α rhythm) of EEG are
   (A) < 4 Hz & 60 mV (B) 4–7 Hz & 40 mV (C) 8–13 Hz & 50 mV
   (D) > 13 Hz & <5 mV

3. How much augmentation is achieved by the augmented EKG leads?
   (A) 50% (B) 75% (C) 100% (D) 200%

4. Which type of laser is not used for soft tissue ablation?
   (A) Nd-YAG (B) He-Ne (C) CO_2 (D) ruby

5. In an ICU (intensive care unit or critical care unit), which instrument is not likely to be present routinely?
   (A) EKG (B) defibrillator (C) dialyzer (D) MEG

Answers

1. D; 2. C; 3. A; 4. B; 5. D

3

Biotelemetry

3.1 Introduction

Telemetry is often essential for long-term monitoring of patients, especially those in intensive care units (ICU) and operation theatres or rooms (OT/OR). It is used during surgery while the patient is under anaesthesia, and also for sleep and epileptic seizure recordings. Special types of electrodes (e.g., floating) are required. Multi-channel data will require multiplexing (frequency or time division) and modulation (amplitude: AM; frequency: FM; pulse amplitude: PAM; pulse position or interval: PPM; pulse code: PCM). The usual modes of transmission through a distance are cable telemetry, telephone line transmission, radiotelemetry and fibre-optic telemetry. Recording of the data and storage may be done on magnetic tapes/cassettes or optical disks or recordable CDs.

Radio (wireless) telemetry allows the examination of physiological data from humans (and animals) in their normal surroundings (away from the laboratory). Telemetric monitoring can be of immense help during the transportation of the patient from home or hospital bed to another ward for checkup or treatment. Two modulators are usually employed for transmitting biomedical signals from afar. A comparatively lower frequency sub-carrier (FM or PWM) is used along with a VHF (very high frequency), which finally transmits the signal (FM) from the transmitter. Double modulation is less noisy in transmitting and receiving low frequency biological signals. The RF (radio frequency) for medical telemetry usually varies from 37–450 MHz.

Telemetry systems may be single channel, multi-channel, multi-patient or implantable.

Presently, medical telemetry systems are available for ECG, EEG, EGG and recordings of temperature, blood pressure and blood flow (BF).

With improved satellite telecommunication in our country, transmission of biomedical signals through satellites is becoming a part of telemedicine/telesurgery systems.

e-health is an umbrella term which includes many functions in its range. It can be defined as 'a new term needed to describe the combined use of electronic (tele)communication and information technology in the health sector' or as 'the use in the health sector of digital data – transmitted, stored and retrieved electronically – for clinical, educational and administrative purposes, both at the local site and at distance'. Further, e-health has also been mentioned as the death of telemedicine, because – in the context of a broad availability of medical information systems that can interconnect and communicate – telemedicine will no longer exist as a specific field. The same could also be said for any other traditional field in medical informatics (Chapter 19), including information systems and electronic patient records. Therefore, e-health presents itself as a common name for all such technological fields. A newer term "tele-health" actually encompasses both 'e-health' and 'telemedicine' (Section 3.3).

3.2 Telemetry

The first step in telemetry, of course, involves the sensor receiving signals. A sample of the sensor signals is then used for transmission and analysis. Sampling takes a snapshot of the sensor signal at discrete times. For most applications, the time interval between samples is kept constant (for example, a sample every millisecond) unless externally clocked.

For most digital converters, sampling is performed by a *sample and hold (S/H)* circuit. An S/H circuit usually consists of a signal buffer followed by an electronic switch connected to a capacitor. The operation of an S/H circuit follows these steps:

i. At a given sampling instant, the switch connects the buffer and capacitor to an input.
ii. The capacitor is charged to the input voltage.
iii. The charge is held until the A/D converter digitises the signal.
iv. For multiple channels connected (multiplexed) to one A/D converter, the previous steps are repeated for each input channel.
v. The entire process is repeated for the next sampling instant.

Hardware can be divided into two main categories based on how signals are sampled: scanning hardware, which samples input signals sequentially, and simultaneous sample and hold (SS/H) hardware, which samples all signals at the same time. The most common hardware used is the Krause® system.

A commercial (Krause®) system comprises:

i. 16-channel sensor inputs (up to 64)
ii. total sampling rate of 80 kHz, simultaneous S+H, 12-bit resolution, 8-pole anti-aliasing filters
iii. acquisition of physical quantities like distances, velocities, accelerations, vibrations, distortions, shocks, revolutions, temperatures, torsions etc.
iv. channel separate signal conditioning with different sample rates
v. parallel online display and data storage to HDD
vi. telemetry and DAT recorder option
vii. extensive software support for signal analysis
viii. 266 MHz Pentium Processor, 32 MB RAM, 4 giga byte HDD (Panasonic® Notebook CF-27)
ix. 12,1 " TFT display with 800 x 600 resolution in high colour (64000 Farben)

The signals from the different sensors will be converted into an analog voltage with a range of ±5V. Each overload is captured and displayed by hardware to support corrections directly on site.

While examining physiological data from animals in their normal surroundings, PIT tags and antenna are common appliances used.

PIT tags are tiny identification chips which are injected into specimens for permanent identification. The chip is read by means of a reader which provides a unique code read out of the chip implanted in the animal. The chip is passive, i.e., does not need any current. The readers can be either portable, working on batteries, or those that require AC. Some have the capability to store codes from many animals before they are downloaded to a computer; others simply display the code of the current reading without storage. In the past, reading distances have been very short (less than 30 cm); however, manufacturers and researchers are adapting readers for more versatile field and laboratory applications.

Care should be taken while using an antenna in biotelemetry applications. Some precautions are listed below:

1. Keep clear of the antenna when taking a bearing. Do not stand, or have assistants stand, within 1/2 wavelength of the antenna elements (1 m at 150 MHz).
2. Protect the antenna elements from being bent out of shape. Bending of elements will alter the pattern and efficiency of an antenna thereby giving misdirecting signals. If the antenna elements get bent, carefully bend them back to their original shape.
3. Periodically inspect the antenna feedpoint to ensure that the connections are still tight and reliable. If the feed apparatus has become loose, have the antenna repaired and re-tuned with proper test equipment.
4. An antenna should be polarised in the direction of the elements in the array. i.e., if yagi elements (yagi is a sharply directional beam antenna and is made up of one driven element and one, or more, parasitic elements) are vertical, the antenna is vertically polarised.
5. Keep all metal objects (including the coax feedline) from interfering with the antenna. The metal elements of an antenna will be influenced by metal objects that are physically parallel to it. With horizontally polarised antennas, it is easy to have the mast and feedline extend away from the antenna perpendicularly. However, in vertically polarised antennas, the mast often has to be attached to the centre of the antenna's boom, right in the middle of the antenna elements. In such a situation, the mast must be non-conducting. The feedline should also loop out behind the antenna and attach to the mast below the antenna elements. If it is not practical to bring the feedline down from behind the antenna, tightly fasten it to the boom and mast with cable ties or use a cantilever design.
6. It should be noted that the horizontal beamwidth of a vertically polarised yagi is greater than the horizontal beamwidth of a horizontally polarised yagi.
7. In animal tracking, using a yagi antenna horizontally helps isolate the direction of the transmitter more precisely; however, when simply trying to find a transmitter, it is wise to try both vertical and horizontal polarisations since the transmitter may be in either orientation.
8. It should be remembered that there is at least a 20 dB loss between polarisations. i.e., when listening to a vertically polarised signal with a horizontally polarised antenna.
9. In fisheries tracking, always use the antenna in vertical polarisation. Horizontally polarised signals experience much greater loss at the air–water interface.
10. Be aware of the sidelobes and backlobe on the antenna. It would not be right to assume with the first peak you see in the received signal that the antenna is pointed at the transmitter. A quick 360-degree scan will ensure that you are checking the greatest peak.

3.2.1 Cable

Nowadays, conventional cable telemetry is hardly used. It may be applied within the hospital itself, but not beyond. However, fibre-optic cables are being laid over many areas and can reliably transmit good quality signals through great distances. Fibre-optic cables are discussed in Section 3.2.4.

3.2.2 Telephone Line

For transmitting biomedical signals through either the public-switched telephone network (PSTN) or plain old telephone system (POTS) telephone lines, an FM reactance modulator, with a central frequency of about 1.5 kHz, is used. This is modulated by ± 200 Hz for a 1 V peak-to-peak signal (linear deviation of about 1%). The demodulator comprises an audio amplifier, a carrier rejection filter and a low pass integrator output circuit to recover the input signal. Coupling or isolation transducers are used at both the transmitter and receiver ends to match the standard telephone line impedance. The data acquisition system (Chapter 7), after line noise correction, converts the analog signals to digital signals through a 12-bit analog–digital converter (ADC), and sends it to the modulator–demodulator (modem), which in turn interfaces with the receiving computer system. However, presently, apart from dedicated leased lines (DLL), Integrated Services Digital Network (ISDN) telephone lines are also being purveyed by Bharat Sanchar Nigam Limited (BSNL). While the PSTN has a bandwidth of 64 kbps (that too theoretically), ISDN lines have a bandwidth of 128 kbps and three of them may even be pooled together to get a 384-kbps connectivity, suitable for uninterrupted, good quality teleconferencing through video.

3.2.3 Radio Frequency (RF)

While using radio frequency systems for transmitting biomedical signals, FM sub-carriers and RF carriers (FM/FM) systems are preferred because of reduced interference. Commonly ECG and recordings of blood pressure and temperature are sent through these systems. However, other modalities (like EGG [2]) are also being tried experimentally. Medically, 2 GHz is the standard frequency used.

3.2.4 Fibre Optics

A device, system or method for continuously and simultaneously monitoring multiple physiological parameters (MMPP, Delphion) from a patient has now been developed and is available commercially. It comprises a precordial (lying on the skin of the chest overlying the heart) strip-patch having a multi-layer flexible structure which permits the telemetering of data by radio frequency, single wire or fibre optics to a hardware recording and display monitor. A plurality of conductive contact elements (CCEs) and microsensors are mounted far apart on the said strip-patch. This permits the simultaneous and continuous detection, microprocessing and transmission of physiological data for monitoring standard 12-lead ECG, cardiac output, respiration rate, peripheral blood oximetry, temperature of a patient, and electrocardiographic fetal heart.

3.3 Telemedicine

Presently, the concept of telehealth comprises telemedicine (remote medical practice) and e-health (healthcare delivery information interchange using the internet)

Telemedicine involves the interconnection of multiple end users (medical and paramedical or health care personnel), through telephone/satellite and/or data network, for sharing/exchanging information in real time. Here, voice teleconference can be enhanced by adding terminals for exchange of text and graphics (telewriting) and also of pictures (video teleconference). For teletransmission of various medical images, various image compression techniques like wavelets or IFS-GA may be used (Section 7.6). Teletransmission may be real-time (online) or off-line (stored and forwarded). If the bandwidth and connectivity is inadequate, online bilateral transmission may not be feasible.

Telemedicine is a rather broad term encompassing teleconsultation, telemonitoring, telediagnosis, teletreatment, and telecare-provision. Essentially it requires good teamwork between healthcare delivery personnel and technocrats.

However, some standards are essential for its wide use. The necessity for standards may be given by the acronym: ICARAS (inter-operability, compatibility, affordability, reliability, accessibility and scalability). In the case of telemedicine, the core standard is the Health Level 7 (HL7, *http: // hl7.org*) developed by the National Committee on Vital and Health Statistics (NCVHS) of the USA.

3.3.1 Standards in Telemedicine

NCVHS has studied the 'issues related to the adoption of uniform data standards for patient medical record information (PMRI) and the electronic exchange of such information'. It recommends that HHS recognise the HL7 (HL7 v 2.2, v 2.3, v 2.4, and later v 2.x) versions and transaction sets as the current PMRI format standard.

Health level 7 includes standards for the following transaction sets:

(i) order entry
(ii) scheduling
(iii) medical record/image management
(iv) patient administration
(v) observation reporting
(vi) financial management
(vii) patient care

HHS recognition of the HL7 versions as current standards means that vendors and users of these versions will not be asked to migrate to newer versions until the more advanced version is fully implementable with supporting implementation guides and conformance tests.

HL7 derives its name from its seven levels:

1. The physical layer deals with data at a bit level.
2. The data link layer breaks input data into data frames and the receiver returns acknowledgment frames.
3. The network layer controls the transmission of packets of data, including routing and control of traffic congestion.
4. The transport layer manages data from the session layer, if necessary splitting it into smaller sections.
5. The session layer allows machines to communicate; this includes synchronization of activity.
6. The presentation layer manages the syntax and semantics of information; this may also include data compression and encryption.
7. The application layer defines file structure and transfer, and manages compatibility between different systems.

For telemedical purposes, a reference information model (RIM) is constructed to facilitate consistent sharing and usage of data across multiple 'local' contexts. In general, the broader the scope of interest, the more important it is to make explicit all assumptions about a topic or domain-of-interest. The HL7 (v 3) RIM is designed to provide a unified framework for, and to serve as a comprehensive source of, all information used by an HL7 specification. The RIM, specifically and unambiguously, articulates both the explicit definitions of healthcare concepts – the 'things of interest' to the world of healthcare information systems – and the relationships (also called 'associations') between these concepts.

HL7 (v 3) specifications (e.g., HL7 v 3 messages, structured documents, etc.) permit loosely-coupled information systems to inter-operate (i.e. exchange data) in a variety of healthcare delivery contexts including those found in disparate provider organisations, perspectives, and jurisdictions. The scope of the HL7 RIM therefore includes all the information that is required to be sent between, and processed by, participating healthcare information systems. In addition, it should be noted that the RIM does not model (nor need to model) information stored by a given healthcare information system which is never transmitted to another system.

Besides the HL7 standard, other important and common standards are:

(i) Digital imaging and communications in medicine (DICOM): The standard which supports retrieval of information from imaging devices/equipment to diagnostic and review workstations, and to short-term and long-term storage systems.

(ii) NCPDP SCRIPT: The standard that communicates prescription information between prescribers and pharmacies. These transactions include new prescriptions, prescription refill requests, prescription fill status notifications, and cancellation notifications.

(iii) IEEE 1073 (1.1.1, 1.2.1, 1.3, 2.1.1, 3.2): A set of medical device communications standards also known as ISO 11073 standards. These standards communicate patient data from medical devices typically found in acute- and chronic-care environments (e.g., patient monitors, ventilators and infusion pumps).

(d) ICD-10 (International Statistical Classification of Diseases, 10th edition, WHO): All the available telemedicine packages do not conform to the standards developed by the World Health Organization (WHO) in the form of ICD-10 for choosing a diagnosis. Therefore, for statistical and epidemiological analyses, it is often difficult to compare medical databases of various countries and the many regions within the same country (especially a big one like India).

The main difficulty in information exchange between two health care centres is the tension between 'local specialisation' and 'global generalisation'. Instead of expecting everyone to think the way we think, it is more user-friendly if all agree to some common semantics.

For telemedical purposes, patient information is recorded in three levels (patient record architecture, PRA)—in Level I, the coded header forms the core; in Level II, the coded context gives more information; while in Level III, the coded content is stored and displayed. The PRA is usually granular. Designing the granularity of any content can sometimes be problematic. Authors typically like content very granular so they know exactly what to put into an element (e.g., overview, procedure step). Very granular content usually results from more semantic models (models with tags that indicate the meaning of the element such as 'overview' instead of tags with generic names such as 'body' or 'para'). *Granules*, itself, are small, efficient components ideal for use in a distributed environment. Small components are faster, require less time to develop and test and can be used throughout the application in endless combinations. This is the basis of appropriate PRA (patient record architecture).

Figure 3.1 shows an oversimplified sketch of a telemedicine setup. All records (text, audio, video and still images) have to be properly coded (compressed if they are very large) before transmitting. While receiving the records, they have to be decoded (and decompressed, if necessary) before reading, interpreting and intervening if necessary.

Whether expensive digital equipment will be easily and affordably available or not at remote rural centres in India is a limiting question. Compared to these digital equipment, analog machines are cheaper but digitisation is necessary for teletransmission. The Health Information Systems (HIS) at both ends may be connected to these sorts of equipment wherever available. A videoconferencing unit at both ends will give a face-to-face consultation environment.

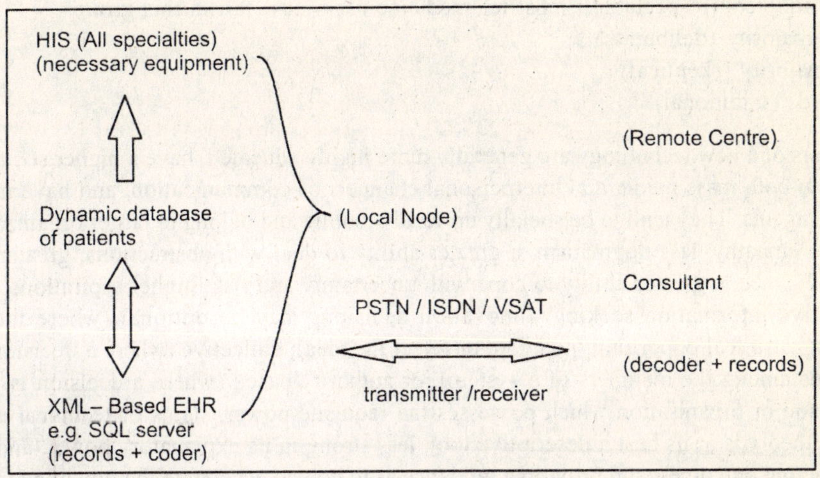

Fig. 3.1 A simplified diagram of a telemedicine setup

3.3.2 Evaluation of a Telemedicine System

A telemedicine system has also to be evaluated to see if it is working properly. Of the various methods available, Rogers' diffusion of innovations or DoI (Rogers, E.M. *Diffusion of Innovations*. 3rd edition. New York: The Free Press. 1983) is one of the most widely accepted theories. The DoI theory is concerned with the manner in which a new technological idea/artifact/technique, or a new use of an old one, migrates from creation to use. According to the DoI theory, technological innovation is communicated through particular channels, over time, among the members of a social system.

Important characteristics of an innovation include:

- relative advantage (the degree to which it is perceived to be better than that which it supersedes)
- compatibility (consistency with existing values, past experiences and needs)
- complexity (difficulty of understanding and use)
- trialability (the degree to which it can be experimented with on a limited basis)
- observability (the visibility of its results)

The stages through which a technological innovation passes are:

- knowledge (exposure to its existence, and understanding of its functions)
- persuasion (the forming of a favourable attitude to it)

- decision (commitment to its adoption)
- implementation (putting it to use)
- confirmation (reinforcement based on positive outcomes from it)

The members of the social system can be categorised according to how and when they adopt the new technology. Different adopter categories are identified as:

- innovators (venturesome)
- early adopters (respectable): most telemedicine users now fall in this group
- early majority (deliberate)
- late majority (skeptical)
- laggards (traditional)

Early users of a new technology are generally more highly educated, have a higher social status, are more open to both mass media and interpersonal channels of communication, and have more contact with change agents. They tend to be socially upwardly mobile and belong to larger organizations. They have greater empathy, less dogmatism, a greater ability to deal with abstractions, greater rationality, greater intelligence, a greater ability to cope with uncertainty and risk, higher aspirations, and engage in more active information seeking. Innovation decisions may be optional (where the person or organisation has a real opportunity to adopt or reject the idea), collective (where a decision is reached by consensus among the members of a system), or authority-based (where a decision is imposed by another person or organisation which possesses the requisite power, status or technical expertise).

The DoI theory is at its best a descriptive tool, less strong in its explanatory power, and less useful still in predicting outcomes, and providing guidance as to how to accelerate the rate of adoption. There is doubt about the extent to which it can give rise to readily refutable hypotheses. Many of its elements may be specific to the culture and time in which it was derived (viz. North America in the 1950s and 60s), and hence less relevant in, for example, East Asian and African countries, and as time goes on.

As far as adoption of telemedicine in our country is concerned, albeit telemedicine has been available for some years, in its current form it remains at a stage of 'early adoption'. Identifying key individuals who will act as early adopters (or who may conversely block an innovation) is thus the first stage of study. Interviews and observation of the meetings that occur during system adoption and the implementation process can provide important evaluation information on shifts in control and the negotiations that occur between the respective groups in India.

3.3.3 Telemedicine Laws in India (see *www.mit.gov.in/telemedicine*)

Regarding the proposed telemedicine law in India, there are quite a few causes for worry. The 'digital divide factor' is a major impediment. Will the new legislation increase the divide between the digital haves and digital have-nots? If so, how will we bridge this gap so that social tensions do not rise to a level of disturbing the society? These are essential factors to be considered.

India has a number of big cities which are well equipped with medical facilities. These cities have high bandwidth communication facilities and corporate hospitals which have the resources to buy hardware and software to conduct telemedicine. However, the need for experts to be consulted online is not critical in these cities since experts in every discipline are available locally and if critical, visiting professionals are available within a short period of time. Documents can also be moved around physically without much delay. Conventional consultancy arrangements, therefore, take care of all the critical needs of patients and telemedical consultancy is not critical except if consulting a foreign expert is required. In contrast, lack of medical expertise is felt more in the rural India where qualified doctors may not be available and transport systems are generally inadequate. The focus of telemedicine in India should therefore be to deliver better quality health services to the rural sector. This may mean that the services should be available at places with minimum telecommunication facilities and low end equipment. To be more precise, it is not a crying need that a CT scan report taken in Mumbai should be forwarded to a super-expert in Delhi online. Rather, it is more important to get an X-ray from a village to reach a mid-level expert in a nearby city for immediate instructions before the patient reaches better facilities.

Presently, there are no clear indications in the proposed guidelines that it has recognised the needs of the Indian rural sector for better health delivery. For example, the guidelines indicate that the 'minimum' bandwidth required for communication should be 128 kbps with 95 % uptime (If an equipment (or system) is able to operate when needed 95% of the time, it is supposed to have 95% 'uptime' (e.g., restarting after a shutdown, either due to power failure or maintenance). Usually the completed product passes the final test 99% of the time, with the result that a rectification and rework station is located only off-line to trouble shoot and (in some cases) repair the failures.). As already discussed in Section 3.2.2, this bandwidth would be available only if a dedicated lease or ISDN line is used. Hence the service is of use only to well-equipped hospitals and not smaller rural medical centres or doctors working in remote places.

If these doctors have to take advantage of teleconsultancy, they should be able to use a normal web cam over a telephone modem to talk/video conference with the consultant and if possible show him a few pictures from a normal digital camera. Obviously the quality of these pictures may fall short of expectations. But still, for the patient and the local doctor, this is a better way to consult than to just speak over the phone. There is a danger that the proposed telemedical guideline may not facilitate such interaction but rather make such interactions illegal. The technical standards suggested in the guideline will require new telemedicine consultation centres to be developed which are equipped with the necessary communication facility, video conferencing facility and the necessary equipment. The cost of such centres would be higher than the cost of setting up a well-equipped diagnostic centre. It is likely that such centres would not be feasible as independent ventures in small towns. They may be viable only in cities or as part of a large hospital.

The guidelines in its present form will only help equipment manufacturers from abroad to sell high cost equipment to the Indian health industry as a replacement for the non-compliant equipments that they may be using now. In actual reality, the only security required is to ensure that every transmission is digitally signed and encrypted with the use of digital certificates. The technical approach of the guideline should therefore be to use ordinary equipment at one end and appropriate applications at the consultant's end. If this is not done, the guideline may widen the digital divide, increase the cost of even the existing health care both in rural areas and in cities and place legal hurdles for rural doctors while consulting experts outside the suggested telemedical system.

In the light of this general trend, progressive medical practitioners and hospitals who are trying to use the developments in information technology to improve the standards of medical service would have to watch their backs, lest they be let down by the very technology they are using. Several reputed corporate entities are putting their reputation on the block through bold initiatives to run 'health portals'. While such initiatives are welcome in the larger interest of the community, it is extremely important for the medical community to realise that there are 'security issues' and the consequential legal liabilities to be dealt with. While the liability for 'wrong' or 'inefficient advice' may be attributed to the medical personnel involved, there is a danger of these people being victimised by either playful or criminally-intentioned hackers who intercept and alter the critical information that may be floating around.

There is always the possibility of an e-mail interceptor changing the dosage of a drug or the drug itself while the doctor's prescription is being transmitted on the net. Similarly, the patient's record in the hospital database can be tampered with, to mislead the doctor attending on him. While this may sound to be a plot of a sci-fi movie at this point of time, one cannot rule out the possibilities of well-planned murder conspiracies with the misuse of technology.

The passing of the Health Information Portability and Accountability Act (HIPAA) 1996, effective from April 2003, in USA was a major policy initiative which had wide ramifications in the health industry. The impact of the act was also felt in the IT industry since it became a key instrument in ensuring compliance through appropriate software support. The proposed telemedicine guidelines which may become a law in due course of time could become the Indian version of the HIPAA since it incorporates the key provisions such as 'universal indicators' and 'privacy protection', although the HIPAA as a comprehensive legislation is more detailed and covers the penal provisions too.

The main motivation behind the HIPAA in USA was to facilitate the smooth functioning of the Health Insurance Scheme. The Act therefore proceeded to mandate 'universal identifiers' for patients, health intermediaries etc. Authentication of different authorities such as the doctor, the dispensary owner etc and the consent of the patient were therefore made an integral part of the system so that disputes in insurance claims could be settled properly. Since there is in existence in the USA, a national Social Security Number to identify every citizen, it was used as the key identifier for the patient and to track the transactions.

Identifiers under telemedicine guidelines

The telemedicine guidelines of India suggest the following three types of identifiers to monitor the system: (i) health care provider identifier, (ii) doctor identifier and (iii) universal patient identifier.

Several advantages are foreseen in such identifier systems. It may avoid duplication of bills, prevent fraudulent health care operators, help consolidation of information, and the management of licenses. In the absence of a Social Security Number in India, it is however going to be difficult to map a patient identifier to any national identity base. The passport, PAN, driving license are all special identifiers and not universal. Hence the Patient Identifier Number, if introduced can become a major exercise of providing a fundamental identification number to every citizen of India who would register himself for telemedical services. This could be the Social Security Identification Number by itself. Since this could later have linkages to issue of death certificates, settlement of insurance claims etc, there has to be an appropriate registration system for this purpose which would avoid duplication. The system of allocating, and managing the patient identifier system, can therefore be the biggest task under the proposed laws. No details have however been given in this regard in the guideline.

The doctor's identifier as well as the health provider identifier would incorporate the license for telemedical practice/consultancy/business and has to be linked to the Registered Medical Practitioner License or such other licenses already monitored by the medical profession. It would be necessary for every medical practitioner under the system to carry a digital certificate so that all his communication is authenticated and encrypted. In view of this, a special digital certificate which incorporates the attributes such as the license number etc has to be made available by the certifying authorities. Handling of such certificates will require a special Certification Practice Statement (CPS) to be drafted. Perhaps just as the IDRBT and NIC are trying to develop themselves into sector-specific certifying authorities, there may be a need for the Medical Council of India (MCI) to develop a separate certifying authority to take care of the digital certificate needs of the telemedicine industry.

Privacy protection under telemedicine guidelines

In order to ensure confidentiality of information, any data either under storage or in transmission has to be digitally signed and encrypted and every user of the system as well as the system itself has to have an inherent capability to affix digital signatures. In case data is being transmitted directly from diagnostic equipment, either in the form of text or static picture or video, it will also have to be encrypted and digitally signed. The present equipment and embedded software in the equipment used in India need to be upgraded to incorporate communication capabilities as well as digital signature capabilities. A CT scan report sent by the scanner will therefore have to be digitally signed by the scanner and delivered through the modem to the network communication channel for onward transmission on the internet or a virtual private network (VPN).

Real-time communication, therefore, may not be a viable proposition in the immediate future. The guidelines have to accept this constraint and avoid imposing standards that are not practically relevant for the time being. In comparison, a 'store and forward' system is simpler and technically and commercially more feasible in the short run.

Any equipment using communication tools for transmission of data will have to be adequately secured against hacking and virus which may cause loss of life. From the security point of view, it would be better to segregate the principal medical diagnostic equipment from the communication devices and prevent any form of external access to the system. The output from the systems should be invariably delivered in a digitally signed electronic form or in print form which would be authenticated by the operator as per the provisions of the Amendments to the Indian Evidence Act (ITA 2000). Even the stored data in electronic form has to be secure against manipulations and the only way this can be done is by having the digital signatures of the doctor/operator/store-in-charge. Each time the data is retrieved, the signatures have to be verified.

The guidelines repeatedly mention that patient's consent would be required in several cases including allowing people for video sessions etc. It must however be remembered that telehealth consultancy would be used mostly in cases where the patient is in critical condition and sometimes even when none of the kith and kin of the patient is available. Hence it would not be possible to get the consent as desired. It may therefore be left to the discretion of the attending doctor to act on behalf of the patient and take all reasonable steps necessary to save the patient's life including waiving the privacy rights of the individual. Perhaps the system may introduce a certification from the attending doctor to this respect as a replacement of the patient's consent.

Another grey area in the guideline concerns the settlement of professional charges between different consulting agencies and in particular the ethical issue of whether a medical practitioner can recommend diagnostic tests and consultancy from a specific source which may amount to 'commercial canvassing'. The medical profession has to sort out this issue since it would be necessary to have service providers in the telemedicine area different from doctors and there cannot be more than a handful of such service providers in any one location. The situation will be that the patient has to avail the services of only the available service provider if the doctor recommends that the diagnosis has to be done in a manner compatible for telemedical consultancy.

The guidelines mention the need to ensure inter-operability of data and information between different users of the system. This essentially boils down to adherence to standards which will perhaps not be dictated by some of the leading equipment manufacturers today. It should be ensured that the law is not misused to create monopolies in the hardware and software required for the system. Inter-operability is a desirable feature which should be sorted out by the equipment dealers and the consumers. The consumers have to be educated enough to demand universal standards and not tie themselves into vendor-specific standards.

It is necessary that the telemedical laws, if enacted, be simple and only provide for recognition of telemedical participants, mandatory use of digital signatures as a means of authentication and an impetus to the vendors of hardware and software to provide open source and IPR unrestricted systems.

In the long run, it is necessary to remind the public that the Government of India has placed the draft guidelines before us and is seeking our comments and suggestions. It is up to us to use this opportunity and participate in the process of legislation. If we remain silent now, we will lose the right to question the laws after they are passed.

Likely applications of telemedicine in India

1. Electronic management and transport of patient information and records for diagnostic purposes.
2. Image compression for efficient storage and retrieval of image data.
3. Image processing for diagnostic purposes.
4. Electronic processing of health and medical claims.
5. Electronic inventory to support community health care organisations.

ISRO (http://www.indiantelevision.com/headlines/y2k5/mar/mar203.htm) plans to launch a communications satellite exclusively for healthcare so that patients and doctors in remote rural areas can consult specialists in cities. Healthsat, to be launched within four years, will have wireless and terrestrial communication links. According to Dr. G Madhavan Nair, ISRO, 'Right now we will use the existing satellite transponders to serve 100 telehealth stations, including district hospitals and speciality hospitals, around the country. Next year we may touch 500 such stations and in 2007 about 1,000. So within the next three to four years we will have one exclusive satellite for the purpose of telemedicine. ISRO may accept help from private corporations to fund the project. By the second half of this year we will bring in private partners. We are also thinking of helping neighbouring countries such as Sri Lanka, Bhutan and Myanmar on their telemedicine projects'.

However, tele-health is only a platform and telecommunication technology on its own is not the simple solution to health problems in India. Since the technocrats are leading most of such movements, the grass-root level players of the HCDS are not taking it seriously. Besides all this, even the basic amenities like potable water, stringent quality control for food and drugs are still far behind the optimal level, even for a developing nation. This must be tackled before dealing with tele-medicine.

3.4 Virtual Reality (VR)

Virtual reality (VR) or a 'simulated environment' purveys an interactive, multi-sensory, virtual world with a three-dimensional computer-generated environment that looks, sounds and feels 'real'. In other words, a subject is 'immersed' in a world of computer-generated, dynamically changing images which allows the user to navigate through the virtual environment. It, therefore, provides the feeling of being one with the environoment—touching, feeling, sensing it with the added advantage of being able to actively modify it. VR is now used for medical training (interactive question–answer sessions), surgical pre-planning (manipulation of images, endoscopic simulation) and also for telesurgical robotic manipulations. Presently it is no more science-fiction that the surgeon be literally continents apart from the patient and yet performs a successful operation. VR is an 'immersive environment' unlike the 'through the window' or non-immersive ones like the computer monitor or the wall screen. It purveys the first opportunity to combine 3-D visual imagery with interactivity at a level that would permit realistic simulation of complex anatomic dissection or performance of surgical procedures. A simulator (e.g., a virtual abdomen) could be connected to a telepresence surgery system. This could be networked and multiple surgeons (trainees) could practise (or innovate) in the computer-generated virtual environment without being exposed to the risk of AIDS or HBV or causing damage to any tissue or organ of the patient. Human interface technology (HIT) is a method to enhance a person's (e.g., a surgeon's) abilities while performing a task, which requires a tool or synthetic device to accomplish it. However, with sophisticated gadgetry taking the upper hand, the 'human touch' should not be overlooked or forgotten.

For impaired cognition, unless the brain is almost totally (more than 90 %) damaged, VR linked to multimedia and hypermedia may be able to 'retrain' some aspects of 'learning'. New concepts and practicable methods of self-care have to be 're-learnt' by paralysed or demented bedridden people. Also, in cases of paraplegics (people with paralysis of the lower half of the body) and quadriplegics (people with paralysis of all four limbs), use of voice or face muscles to move the wheelchair by oneself, may be easily learnt with the help of VR. The movements of facial muscles or sounds produced by the voice have to be pre-amplified and sent to the panel controlling the direction and speed of the movement of the wheelchair. Unique commands should have predetermined results. These can be modelled through VR and learnt easily without injuring oneself.

For impaired homeostasis (Section 16.2), self-checking, again through AI programs linked to VR, multimedia and hypermedia may be able to reduce the risk of an impending failure of any organ. If emergency measures can be instituted, sufficiently in time, a lot of complications can easily be avoided.

There is a great need for more academicians with a thorough understanding of information technology, quantitative methodology, medical care and the socioeconomic environment in which medical care is and will be practiced. In our country, the priorities of health care delivery systems are not well-defined. However, with better telecommunication facilities being available everyday, medical informatics is likely to become cost-effective under most of the conditions prevailing in India. A better link between advanced cities and backward villages will ultimately benefit the ailing commoner of India. From this viewpoint, the future of medical informatics seems bright in India, provided, positive efforts are made in this direction.

3.5 Telesurgery

Microsurgical telerobot systems have been developed according to the results of operation task analysis. Several fields of microsurgery have been investigated and surgery tasks and tool motions have been analysed to acquire guidelines for specifications of a microsurgical telerobot system. For a slave robot system, a 6-dof (degree of freedom, see Section 15.3) parallel micro-manipulator, which is of a modified Stewart platform type, has been developed and attached to the industrial robot for macro-motion. A 6-dof force-reflecting master device which allows a user to touch, feel, manipulate, create, and/or alter simulated objects in a virtual environment, has also been developed using five-bar parallel mechanisms driven by harmonic DC servomotors. A bilateral control algorithm has been proposed based on the force-position loop and the local position controller for the master device eliminates the effect of gravity, friction, and inertia.

With the help of the internet and telecommunications technology, a doctor in Baltimore can operate on patients all over the world without leaving his home office. Using a PC and four ISDN lines, Dr Louis Kavoussi of Johns Hopkins Bayview Medical Center controls robotic surgical tools and cameras remotely and can transmit and view images in real time. During surgery, Kavoussi can view either the operating room or inside the patient. He can also give surgeons written assistance and operate a device that burns and seals tissue as well as control robots that hold cameras or place needles in the patient's body. Kavoussi says, 'Our applications have been used specifically for what's called minimal invasive surgery, examples of that are laparoscopies, putting a little tube in the stomach to look around, arthroscopy, looking at knee joints, thoracoscopy, looking at the chest, and angoscopy, looking at blood vessels. These techniques involve insertion of instruments and viewing equipment into the body through natural orifices or punctures created by the surgeon. The main advantage is rapid patient recovery, but the disadvantage is associated with reduced dexterity and workspace, and the sensory input to the surgeon being available only through a single video image. With the help of high-speed data lines and advanced robotics, surgeons will eventually be able to perform and complete operations remotely from anywhere in the world. There is no doubt in my mind that this is the way surgical care is performed in the future'.

Laparoscopy

In order to facilitate a virtual reality simulation of laparoscopy, based on a realistic description of the interaction between the target tissues to be treated and the surgical instruments used for an operation, for training purposes, a FE model of the involved objects is needed which is capable of running in real time. For this purpose, a stable and explicit FE formulation for the integration of the time-dependent continuum-mechanical equations is being developed. A highly parallel implementation on a custom-built multiprocessor system is envisaged in order to satisfy the enormous speed requirements associated with a real-time operation.

US surgeons have taken telemedicine one step further when they used a dedicated fibre-optic line and a remote-control robot to remove the gall bladder of a patient in an operating room in France—more than 4,000 miles away. This was achieved by Professor Jacques Marescaux and his 40-member team from the Institute for Research into Cancer of the Digestive System (IRCAD). On September 7, 2001, working from New York, Marescaux removed the gall bladder from a 68-year-old woman in Stroudsburg, France. It was the first major, trans-Atlantic telesurgical operation ever to be completed. The patient suffered no complications and returned home two days later.

In New York Marescaux's team, using a video screen and controllers connected with high-speed fibre-optic lines, directed the arms of a Zeus laparoscopic surgical robot in France, 4,300 miles away. Doctors at the European Institute of Telesurgery in Stasbourg watched on during the 54-minute operation, ready to step in should a problem arise. As the robot was manipulated, the surgeon watched the machine's responses to his hand motions on a video screen. The time delay between his movements in NY, action in France and the visualization of that movement on the NY screen was only *155 milliseconds*, virtually imperceptible to the human eye. That tiny time delay was the key to the success of the procedure. Experiments on pigs showed that a delay of more than 330 milliseconds would be too risky to the patient.

France Telecom made the high-speed connection possible. The optimised compression developed by the company allowed video coding and decoding to be lightning fast. Combined with quick signal transmission, researchers were able to provide an even smaller delay than required. 'This was a tremendously challenging project, since we had to apply our expertise in high-speed services within an extremely exacting environment in terms of the reliability and security needed,' Jean-Pierre Temime, Director of Marketing for France Telecom Enterprise Services, said.

Of course, the connection was not the only innovation. A special laparoscopic robot and complex software were developed by the engineering staff at Computer Motion, a high-tech company that develops medical equipment. They hope the new SocratesTM Tellecollaborative System, and MicrowristTM technology for the ZeusTM Surgical Robotic System will usher in a time of increased 'tele-enabled surgery and real-time collaboration ... We have taken an enormous step... It's a good beginning to a new era,' said Dr. Maky Zanganeh, Chairman and Chief Operating Officer for Computer Motion in Europe. All told, the operation cost $1 million. The experiment was described in *Nature*. 'It's a phenomenal step and we can't even begin to imagine the implications for medicine... the barriers of space and distance have collapsed,' explained Marescaux.

Telesurgery may also be employed during future space exploration. Travelling to Mars and back may take three or more years, during which time astronauts may need access to medical and surgical care.

The combination of robot, video and telecommunication technologies will make it possible for surgeons to perform operations on patients, anywhere on the planet. Applied telesurgery may make it possible to allow specialists to aid surgeons in war zones or train doctors in developing countries. Marescaux stated, 'It lays the foundations for the globalisation of surgical procedures, making it possible to imagine that a surgeon could perform an operation on a patient anywhere in the world.' or train doctors in developing countries.

3.6 Further Reading

[1] Raja Rao, C and Guha, S.K. *Principles of Medical Electronics and Biomedical Instrumentation.* Universities Press Pvt. Ltd. 2001.

[2] Gacsalyi, U, Zabielski, R and Pierzynowski, S.G. Telemetry facilitates long term recording of gastrointestinal myoelectrical activity in pigs. *Exp. Physiol.* 85: 239–241. 2000.

[3] Mitchell, J. Increasing the cost-effectiveness of telemedicine by embracing e-health. *J Telemed Telecare* 6 Suppl 1:S16-9. 2000.

[4] Let your dreams come true through virtual reality (Current Topic). *J. Indian Med. Assoc.* 95: 190. 1997.

[5] Sarbadhikari, S N. Will the aging computer technology comfort the aging Humans? *Speculations in Science & Technology* (Special Issue on Computers in the Fight against Aging) 19(2): 131–136. 1996.

[6] Gotman, J., Ives, J.R. and Gloor, P. Quesney, L.F., Bergsma, P. Monitoring at the Montreal Neurological Institute. *Electroencephalogr Clin Neurophysiol Suppl.* 37: 327-40. 1985.

[7] Maheu, M, Whitten, P and Allen, A; *E-Health, Telehealth, and Telemedicine: A Guide to Startup and Success.* Jossey-Bass. 2001.

[8] Darkins, A.W and Cary, M.A. *Telemedicine and Telehealth: Principles, Policies, Performance and Pitfalls.* Springer. 2000.

[9] Norris, A.C.; *Essentials of Telemedicine and Telecare.* John Wiley and Sons. 2002

[10] Carter, J.H. *Electronic Medical Records: A Guide for Clinicians and Administrators.* American College of Physicians. 2001

[11] Coiera, E. *A Guide to Medical Informatics, the Internet and Telemedicine.* OUP. 1997: *www.coiera.com/*

[12] WebSurg: Operation Lindburgh

[13] www.Naavi.com

[14] www.jmir.org

[15] *http://www.noie.gov.au/projects/ecommerce/ehealth/rise_of_ehealth/unstoppable_rise.htm* http://www.newscientist.com/article.ns?id=dn1316, http://www.1800calldoc.com/newtecharticles/telesurgery.htm, http://www.hoise.com/vmw/01/articles/vmw/LV-VM-10-01-20.html

3.7 Questions

1 Descriptive Questions

1. What is telemetry? Presently what tests are available for medical telemetry systems?
2. What are the prerequisites and applications of telemedicine ?

2 Write Short Notes on

1. FM sub-carrier
2. RF for medical telemetry
3. Virtual reality
4. Telesurgery
5. Double modulation

3 Multiple Choice Questions:

Choose the single best response.

1. The term e-health encompasses, in the health sector, all except
   (A) electronic (tele)communication (B) information technology
   (C) the use of digital data for clinical, educational and administrative purposes
   (D) only at a distance.

2. A virtual reality environment is
   (A) immersive (B) one-way (C) not widely applicable (D) not to be used as an educational tool

3. Biotelemetry systems are useful for
   (A) home care (B) ambulant patients (C) patient located at remote inaccessible places (D) all

4. A telemedicine link should not be through
   (A) ISDN (B) courier (C) VSAT (D) PSTN

5. The medical records data (including images and signals) may be transmitted
   (A) interactively (B) store and forward (C) either (D) none.

Answers

1. D; 2. A; 3. D; 4. B; 5. C

Section 2

Electrophysiological Signal Analysis and Processing

Chapter 4 deals with the generation, transmission and interaction of signals in the biological nervous system and gives a brief outline of their mathematical models.

In Chapter 5, acquisition of ECG, EEG, EMG, their amplification, digitisation, storing and display are discussed.

Some signal processing tools like fast Fourier transforms (FFT), wavelets and non-linear dynamics (chaos) are described in Chapter 6.

Chapter 7 discusses some methods of medical data acquisition, analysis, compression and monitoring.

4

Interaction of Signals in the Neuro-muscular System and their Models

4.1 Introduction

As described in Chapter 1, electric potentials are generated in the nervous system as local, graded (e.g., generator or receptor, synaptic, end plate) potentials and as propagated, non-graded (all-or-none) action potentials. They are transmitted through nerve and muscle fibres and the junctions of two such fibres are the functional units like synapse (between neurons) or neuromuscular/myoneural junctions (between neuron and muscle fibres). The potential generated at the synapse is called *post-synaptic potential* (PSP), and may be excitatory (EPSP) or inhibitory (IPSP), as described in Fig.4.1.

The interactions between these signals could be excitatory (positive), inhibitory (negative) or oscillatory (varying between positive and negative). The various interactions at each functional unit give rise to several configurations of the total nervous system. The effects are multifarious viz., subliminal fringe, fractionation, occlusion, facilitation, potentiation and depression—the last two being either short-term (ionic or electrical) or long-term (structural changes). There have been various mathematical models designed to estimate and predict the complex interrelations between functional

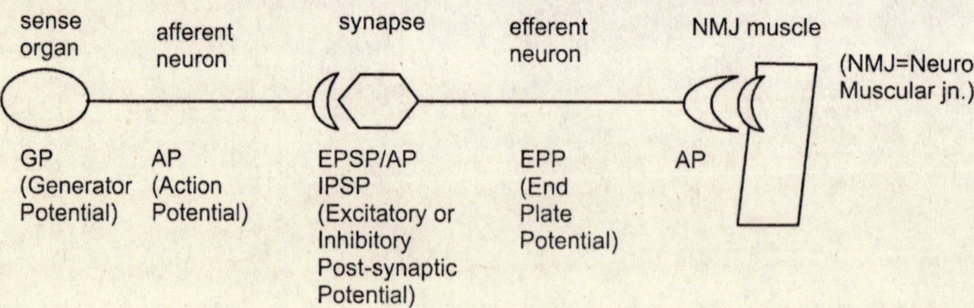

Fig. 4.1 Simple schema of nerves and muscles

units. Earlier, complex linear methods were employed to describe them. Now many theorists are applying non-linear dynamics or chaos theory to describe the interactions between about 10^{14} to 10^{16} interconnections within the human brain.

4.2 Nerve–muscle Interactions

Muscle contraction caused by changes in the neural fibres will be discussed in detail in Section 14.3. For learning and memory of common, familiar movements, consolidation is achieved by structural changes in the dendrites and their connections. These changes can be modified by external influences like conditioning (which may be associative like Pavlovian or operant/Skinnerian, like reward/ punishment). Long-term memory causes activation of genes that enhance synaptic contacts through the interaction of proteins like calpain and fodrin.

Computer modelling and simulation of human movement plays an increasingly important role in sports and rehabilitation, with applications ranging from sport equipment design to understanding pathologic gait. The complex dynamic interactions within the musculoskeletal and neuromuscular systems make analysing human movement with existing experimental techniques difficult but computer modelling and simulation allows for the identification of these complex interactions and the causal relationships between input and output variables. Neptune (2000) provides an overview of computer modelling and simulation and presents an example of an application in the field of rehabilitation.

Progress over the past decade has made it possible to construct a simple model of neurotransmitter release. Currently, some models use artificially formed vesicles to represent synaptic vesicles. A planar lipid bilayer is used to represent a pre-synaptic membrane. Fusion of vesicles with the bilayer is via channel proteins in the vesicle membrane by an osmotic gradient. Woodbury (1999) presents a framework for the successful construction of a more complete model of synaptic transmission. This model includes real synaptic vesicles that fuse with a planar bilayer. The bilayer contains acetylcholine receptor (AChR) channels which function as autoreceptors in the membrane. Vesicle fusion is initiated following a Ca^{2+} flux through voltage-gated Ca^{2+} channels. Key steps in the plan are validated by mathematical modelling. Specifically, the probability that a reconstituted AChR channel opens following the release of ACh from a fusing vesicle, is calculated as a function of time, quantal content, and number of reconstituted AChRs. Experimentally obtainable parameters for construction of a working synapse are given. The inevitable construction of a full working model will mean that the minimal structures necessary for synaptic transmission are identified. This will open the door in determining regulatory and modulatory factors for transmitter release.

The process of neurotransmitter release at the neuromuscular junction needs to be represented appropriately while modelling the synaptic chemical transmission as a reaction–diffusion system. The release mechanisms of the expanding pore of the synaptic vesicle, which releases the neurotransmitters from the end of the neuron and the acceleration of diffusion have been analysed by computer simulation (Naka 1999) with respect to the effects of the characteristic parameters in the mechanisms on spontaneous generation of the miniature end plate current or potential (MEPC or MEPP). It has lead to the following evaluation. In the expanding pore mechanism, an expanding rate of more than 10 nm ms^{-1} and a diffusion coefficient of acetylcholine in the synaptic cleft D_c of about 1.0×10^{-6} cm^2 s^{-1} yield maximum amplitude; the rise time and the decay time constant of the MEPC being in agreement with the empirical data. In the active release mechanism, a 10-fold acceleration of the natural diffusion and a similar value of D_c are required to fit the empirical MEPC

Neuromuscular reflexes with time-delayed negative feedback, such as the pupil light reflex, have different rates depending on the direction of movement. This asymmetry is modelled by an implicit first-order delay differential equation in which the value of the rate constant depends on the direction of movement. Stability analyses are presented for the cases when the rate is (1) an increasing and (2) a decreasing function of the direction of movement. It has been shown (Atay et al. 1998) that the stability of equilibria in these dynamical systems depends on whether the rate constant is a decreasing or increasing function. In particular, when the asymmetry has the shape of an increasing step function, it is possible to have a stability which is independent of the value of the time delay or the steepness (i.e., gain) of the negative feedback.

There is a steadily growing body of experimental data describing the diffusion of acetylcholine in the neuromuscular junction and the subsequent miniature end plate currents produced at the post-synaptic membrane. To gain further insights into the structural features governing synaptic transmission, Smart et al. (1998) have performed calculations using a simplified finite element model of the neuromuscular junction. The diffusing acetylcholine molecules are modelled as a continuum, whose spatial and temporal distribution is governed by the force-free diffusion equation. The finite element method was adopted because of its flexibility in modelling irregular geometries and complex boundary conditions. The resulting simulations are shown to be in accord with experiments and other simulations.

4.3 Neuromuscular Control

Nerve endings are biological transducers converting electrical into chemical energy, by the synthesis, storing and releasing of neurotransmitters. The secreted neurotransmitters act on specific post-synaptic membrane receptors and are rapidly removed from the synaptic cleft by diffusion, metabolism and

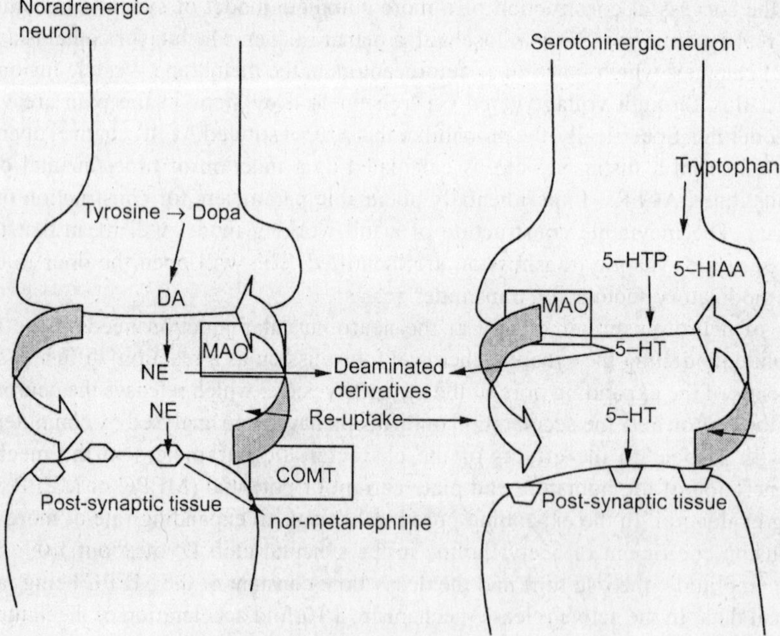

Fig. 4.2 Examples of re-uptake at nerve terminals

often re-uptake by pre-synaptic neurons. All these processes, including the post-receptor events in post-synaptic neurons are controlled by various physiological factors. Pre-synaptic receptors or autoreceptors often inhibit further secretion of ligand purveying negative feedback (e.g., norepinephrine acting on a_2 pre-synaptic receptors inhibits norepinephrine secretion). Moreover, they may also facilitate positive feedback. For receptors, down regulation (decrease in the number of receptors due to prolonged exposure to the specific ligand: negative feedback) and up regulation (increase in the number of receptors or positive feedback) also play important roles. Re-uptake (Fig. 4.2) is a major factor in terminating the action of neurotransmitters. By inhibiting re-uptake, effects of neurotransmitter release are more and prolonged. All these factors can be modelled mathematically.

To cite an example, for the peripheral motor control system, control signals come from interneurons (having force feedback from golgi tendon organs (GTOs) in muscles, which respond to muscle force and load), a motoneurons (motor neurons), dynamic γ motoneurons and static γ motoneurons (the latter 2 responding to length and velocity feedback in muscle spindles). The α motoneurons get feedback from the GTOs (through interneurons) and muscle spindles, and send efferent signals to the motor end plate (in the myoneural junction). The muscle is both the beginning (receptors in muscle spindles or intrafusal fibres and GTOs, responding to changes in muscle length or force) and the end points (ultimately leading to contraction of contractile elements or extrafusal fibres).

The involuntary contraction of a skeletal muscle by stretching, with an intact nerve supply, is known as the *stretch reflex* or the *deep tendon reflex (DTR)*. These are mono-synaptic reflexes originating from the muscle spindle, conducted by fast sensory fibres, to the motoneurons supplying the same muscle.

Interestingly, during a stretch reflex, the antagonistic muscles (those opposing its action, e.g., biceps or flexor of arm and triceps or extensor of arm) relax. This phenomenon is caused by the reciprocal innervation of the antagonistic muscles. Impulses in the Ia fibres of the agonist or protagonist muscle (e.g., biceps brachii) lead to post-synaptic inhibition of the motoneurons in the antagonist muscle (e.g., triceps brachii). This is a bi-synaptic pathway, where collateral from each Ia fibre passes in the spinal cord through the inhibitory interneuron, which synapses directly on one of the motoneurons supplying the antagonist.

Another interesting feature of the stretch reflex is that the contraction is greater with a larger force, but only up to a limit. After that, the muscle actually relaxes. This is known as the inverse stretch reflex or autogenic inhibition, mediated by the GTO.

For neuromuscular secretion, stochastic models are commonly used. For instance Bykhovskaia et al. (2000) evaluated two models of neurosecretion in terms of their ability to predict the dependency of quantal content (m) on the frequency of repetitive stimulation of a lobster motoneuron. First, the hypothesis that neurosecretion is limited by a fixed number of release sites was tested by the fit of the distribution of m by uniform and non-uniform binomial statistics. The obtained release probabilities suggested that frequency facilitation can be due to activation of a group of sites with high release probabilities. However, the fit obtained using this model was not statistically significant due to the large number of fitting parameters. Secondly, the hypothesis that neurosecretion is limited by the rates of exchange between the releasable pool and the total store of quanta and that each stimulus enhances quantal mobilisation were tested. Monte Carlo simulation was carried out in accordance with this model and the observed distribution of m with very few fitting parameters and therefore with a high level of significance (>0.1) was reproduced. This result demonstrates that the mobilisation of extra vesicles with each stimulus is a mechanism that allows a very accurate and parsimonious quantitative description of frequency facilitation.

While stochastic models are used for neuromuscular secretions, non-linear dynamics is now becoming the key player in cases like the brain. Tamura and Saito (2002) describe a rheological motor model that satisfies the major mechanical properties of the skeletal muscle. The model consists of two Maxwell elements and a Voigt element connected in parallel with each other and has a force generator in it. The model well explains the mechanical behaviour in the quick and slow recovery phases of a muscles' isometric contraction and how it achieves a sufficient isotonic shortening speed. The energy liberation of the motor in isotonic contraction has been calculated and a mechanism of control proposed, which operates so as to decrease the dissipated energy by altering the weights of the elastic and viscous constants in Maxwell elements. Thereby it becomes possible for the motor to possess non-linearity in energy liberation and load–velocity relation alike in the muscle. The model would serve as a base model to be utilised for analysing the kinetics of human macrosystems and/or for modelling the human neuromuscular system of motion control.

4.4 Mathematical Models

Figure 4.3 shows a diagram of an equivalent circuit of a neuron or nerve cell having a diameter of 1 mm. R_{ext} denotes the resistance in the exterior of the cell (negligible resistivity in the external medium of ECF (extra cellular fluid) between the axon and the bathing medium), while R_{int} denotes the resistance (10^7 W in 1 cm) in the internal axoplasm of the cell. R_m denotes the resistance in the membrane of the cell and is about 10^{10} W in 1 cm of the membrane, relative to the radial current flow. The capacitance C of this neuron is about 1 mF/cm^2. In this section, we shall concentrate more on the interactions between nerves and muscles. For a discussion on brain modelling, please see Section 6.4.1.

Martineau and Gardiner (2002) believe that albeit mechanical forces regulate many cellular processes, mechanotransduction, is poorly understood. In the skeletal muscle, mechanical forces have a major impact on the regulation of cellular volume, yet the nature of the mechanical stimulation to which the muscle is most sensitive is unknown. It was recently demonstrated that the activation of the mechanically-sensitive kinase p54 jun-N-terminal-kinase (JNK), is a quantitative marker of mechanical stimulation in skeletal muscle. This marker was shown to be more sensitive to peak tension than to other tension-related parameters such as the tension–time integral (TTI) and the rate of change of tension (*dT/dt*). The purpose of the study was to find out the contribution of TTI and *dT/dt* to the mechanical stimulation of muscle under conditions of constant peak tension. The experimental rat's medial gastrocnemius *in situ* was subjected to one of four 5-min passive stretch protocols consisting of equal length excursions, but differing in displacement–time integral (4%, 40%, or 100%) and/or rate of stretch (0, 3, or 30mm/s), and the resulting p54-JNK phosphorylation was assessed. A linear

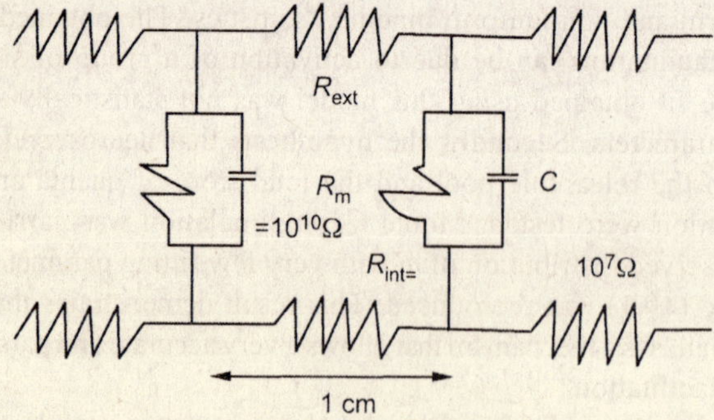

Fig. 4.3 Equivalent circuit of a neuron

relationship between TTI and p54-JNK signalling was observed. However, it was observed that *dT/dt* did not have any significant effect on p54-JNK signalling. It was concluded that peak tension and TTI are necessary parameters for modelling the mechanical stimulus–response of the muscle. Additionally, the mechanism of mechanotransduction is sensitive to peak tension and TTI, but not to *dT/dt*, and thus exhibits spring-like behaviour. These findings may contribute to the refinement of disuse atrophy counter-measures.

Kirsch et al. (2001) argue that neuroprostheses with the use of functional neuromuscular stimulation (FNS) or functional electrical nervous stimulation (FENS) have the potential to restore elbow and shoulder function lost to paralysis because of spinal cord injury (SCI). The human shoulder is highly flexible and thus provides a wide range of motion to the arm and hand, although at the expense of precarious stability of the articulations. The complexity of the shoulder has prevented widespread use of FNS at this joint. However, musculoskeletal modelling of the elbow and shoulder has the potential to significantly accelerate the development of neuroprostheses by allowing many mechanical issues to be resolved in simulation prior to implementation in human subjects. They describe their rationale for the use of musculoskeletal modelling, the model they are using, and several practical applications of the model to study the potential use of shoulder and elbow muscle FNS to restore function following cervical SCI.

Dorgan and Reilly (1999) propose a mathematical model for the bulk electrical impedance of human skin. In particular, their model describes the impedance of skin during surface FNS with square stimulation pulses. Experimental data are presented that illustrate the non-linear dynamic properties of human skin during current and voltage controlled stimulation. Model predictions are compared to experimental data, measured under both constant voltage and constant current transcutaneous stimulation. It is found that this model captures a variety of non-linear time-varying effects observed in the skin impedance when stimulated with either protocol. This model may be used as part of large neuromusculoskeletal models or in the more accurate modelling of transcutaneous FNS, which is currently the most common clinical implementation of FNS.

The objective of this chapter was to give a glimpse of the varied aspects of neuromuscular modelling that is being done in the world at present. If interested, please refer to the list given below.

4.5 Further Reading

[1] Tamura, Y and Saito, M. A rheological motor model for vertebrate skeletal muscle in due consideration of non-linearity. *J Biomech.* Sep 35:1273. 2002.

[2] Martineau, L.C and Gardiner, P. F. Skeletal muscle is sensitive to the tension-time integral but not to the rate of change of tension, as assessed by mechanically induced signaling. *J. Biomech.* 35: 657–63. 2002

[3] Kirsch, R.F et al. Model-based development of neuroprostheses for restoring proximal arm function. *J. Rehabil. Res. Dev.* 38: 619–26. 2001

[4] Ganong, W.F. *Review of Medical Physiology* 20th ed. Prentice Hall Int. (Inc.). 2001.

[5] Guyton, A.C and Hall, J.E. *Textbook of Medical Physiology* 10th ed. W B Saunders. 2001.

[6] Neptune, R.R. Computer modeling and simulation of human movement. Applications in sport and rehabilitation. *Phys. Med. Rehabil. Clin. N Am.* 11: 417–34. 2000.

[7] Bykhovskaia, M, Worden, M. K and Hackett, J.T. Stochastic modeling of facilitated neurosecretion. *J Comput Neurosci.* 8:113–26. 2000.

[8] Dorgan, S. J and Reilly, R. B. A model for human skin impedance during surface functional neuromuscular stimulation. *IEEE Trans. Rehabil. Eng.* 7: 341–8. 1999.

[9] Woodbury, D.J. Building a bilayer model of the neuromuscular synapse. *Cell Biochem. Biophys.* 30: 303–29. 1999.

[10] Naka, T. Evaluation of characteristic parameters for the neurotransmitter release mechanisms at the neuromuscular junction. *Biosystems* 49: 143–9. 1999.

[11] Atay, F.M and Mallet-Paret, J. Modeling reflex asymmetries with implicit delay differential equations. *Bull Math. Biol.* 60: 999–1015. 1998.

[12] Smart, J.L and McCammon, J.A. Analysis of synaptic transmission in the neuromuscular junction using a continuum finite element model, *Biophys. J.* 75: 1679–88. 1998.

4.6 Questions

1. Descriptive Questions

1. How is our musculoskeletal or neuromuscular system controlled?
2. Presently, for modeling neuromuscular activity, which methods are given preference?

2. Write Short Notes on

1. Synapse
2. Neuromuscular or myoneural junction
3. Local potentials
4. EPSP and IPSP
5. Re-uptake
6. Stretch reflex
7. Reciprocal innervation

3. Multiple Choice Questions

Choose the single best response

1. The contractile response in skeletal muscle
   (A) starts after AP is over.
   (B) does not last as long as AP.
   (C) produces more tension when muscle contracts isotonically than when muscle contracts isometrically.
   (D) can increase in magnitude with repeated stimulation.

2. For neuromuscular transmission of impulse which one is not required?
   (A) N_G type of acetylcholine receptor
   (B) calcium ion
   (C) intact motoneurons
   (D) action potential

3. Which statement is incorrect? Unlike a neuromuscular junction, a synapse has
   (A) multiple inputs.
   (B) slower conduction.
   (C) EPP.
   (D) only excitation.
4. Neuromuscular modelling is essential for all except
   (A) sports.
   (B) rehabilitation.
   (C) mathematics.
   (D) understanding better.
5. Modelling may be done by
   (A) stochastic methods.
   (B) nonlinear methods.
   (C) either
   (D) none

Answers

1. D; 2. A; 3. B; 4. C; 5. C

5

Acquisition, Digitisation, Storing and Display of some Important Biomedical Signals

5.1 Introduction

Electrophysiological signal processing deals with the representation of (electrophysiological) signals as ordered sequences of numbers and the processing of these sequences. The most important requirements for it are estimation of characteristic electrophysiological signal parameters (or electrophysiological signal analysis), elimination or reduction of unwanted interference or noise and the transformation of the signal into a more informative form.

5.2 Recorders

For recording biomedical or electrophysiological signals, electrodes, transducers, signal conditioners (pre-amplifier, amplifier, filters—hardware/analog and software/digital) and writing systems (direct ink pen or heat) are necessary.

Ultraviolet (UV), thermal array, electrostatic and light gate array recorders are also available.

Digital recordings on magnetic tapes or disks are now done—either along with analog paper recordings or alone. Recording is done by magnetising the disk (or tape) to saturation in either of its two possible directions at discrete points along its length. The advantages are its high accuracy, simplicity and ability to reproduce electronic circuitry.

For instance, analog EEG machines are steadily being replaced, especially in teaching hospitals abroad, by digital ones with subsequent enhancement in precision for quantification of events. Earlier 4-channel recording of evoked responses (ER) or potentials (EP) are now carried out by full electrode complement where each electrode is assigned to represent the output of one filter in the normal processor. Data reduction methods and high speed desktop computers can enable a doctor to scan a 1½ hour recording in just three minutes. Greater channel capacity and realistic head models based on individuals, is likely to make magnetic resonance image dipole source modelling almost as useful as MEG.

The 8-channel polygraph mentioned in Section 2.5 is an example of a recording device. Another design of a system for electrophysiological signal analysis is presented in Fig. 5.1. The components of this system are described in this chapter. Amplifiers have already been described in Section 1.5. Writing devices of recorders can be categorised as in Table 5.1

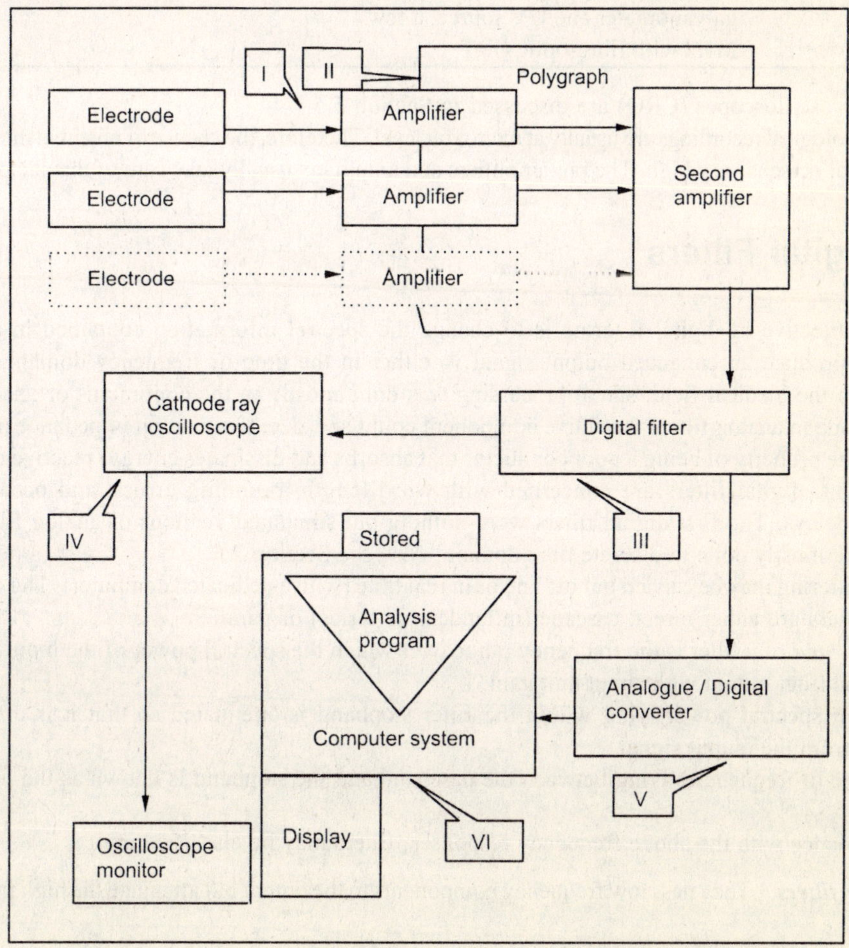

Fig. 5.1 Design of a system for electrophysiological signal analysis

Table 5.1 Comparison between various types of recorders

Type	Frequency response	Recording mode
Potentiometric	1 Hz at 25 cm peak-to-peak or up to 6 Hz at ↓ amplitude	Inexpensive strips of chart paper and ink
Direct-writing ac	60 Hz at 40 mm peak-to-peak or up to 200 Hz at ↓ amplitude	Stylet with heat (on waxed paper) or on plain paper (ink)
Ink jet	Up to 1 kHz	Inkjet on plain paper
Electrostatic	Up to 10 kHz	Electrostatic
Thermal array	20 Hz sine wave is defined by 10 adjacent segments, up to 200Hz	Thermo-sensitive plain paper
Ultra-violet	Up to 2 kHz with mirror galvanometer and UV source, a few MHz with fibre-optic CRT	Special UV-sensitive paper

Cathode ray oscilloscopes (CRO) are discussed in Section 5.5.
Electrophysiological recordings are usually at microvolt level. Therefore, the chance of potential interference with electrical noise is very high. The pre-amplifiers and amplifiers usually take care of this factor.

5.3 Digital Filters

The chief objective of digital filtering is to change the spectral information contained in an input signal x_t to produce an enhanced output signal y_t, either in the time or frequency domain. Earlier, especially in the medical field, signal processing was done mostly in the continuous or analog time domain. While in analog filters the active component count and size, termination impedance matching and lossy (the property of being a poor conductor that absorbs and dissipates energy) reactive elements are important, digital filters are concerned with word length, rounding errors, and occasionally processing delays. The first digital filters were nothing but simulated versions of analog filters, but now they are mostly done in discrete time domain. Also see Section 7.5.

Digital filtering may be carried out off line or in real time (with a dedicated computer). The common forms of filters are either direct, cascade (in tandem or series) or parallel.

The *passband* of a filter is the frequency range over which the spectral power of the input signal is passed to the filter output with about unit gain.

The input spectral power lying within the filter stopband is attenuated so that it is effectively eliminated from the output signal.

The range of frequencies lying between the passband and the stopband is known as the *transition band.*

In accordance with the above frequency responses, filters may be classified into:

(i) ***Low pass filters*** They pass low frequency components to the output but attenuate the high frequency constituents.

(ii) ***High pass filters*** They pass high frequency components to the output but eliminate the low frequency waves.

(iii) ***Band pass filters*** They pass only an intermediate (neither high nor low) range of frequencies specified. It may be realised as a cascade of low pass and high pass filters whose passbands overlap.

(iv) ***Band stop filters*** They act in the opposite manner to band pass filters, passing high and low frequencies, while rejecting intermediate frequencies. This may be achieved by putting low pass and high pass filters with non-overlapping passbands, in parallel.

Table 5.2 Differences between an IIR and an FIR filter

	IIR	FIR
Advantages	Sharp frequency cutoff characteristics can be attained with a relatively low-order structure, so processing time and hardware requirements are reduced	Can have exactly linear phase characteristics, and implementation through fast convolutions lessens the computational burden
Limitations	Lacks exact linear phase characteristics	Requires many coefficients
Applicability	Wide because many common analog filters are easily converted to IIR digital filters	Popular since fast convolution routines are now available easily

Impulse responses (infinite or IIR and finite or FIR) also are used for classifying digital filters. Their differences are tabulated in Table 5.2.

The filter response is usually expressed as the squared magnitude $\left|H(e^{j\omega})\right|^2$. Otherwise, the power gain may be defined in decibels as 10 $\log_{10}\left|H(e^{j\omega})\right|^2$

5.4 Sampling Techniques

Before discussing sampling techniques, it will be useful to classify signals as deterministic and non-deterministic (random). Deterministic signals may be further categorised as periodic, finite-duration, transient and quasi-periodic (almost periodic). Non-deterministic signals are rather unpredictable, and so they are described statistically (e.g., as averages).

For digital recording, a sampling technique is used to measure a varying signal. The sampled readings are converted into a binary code, through an analog-to- digital converter (ADC). For analysis, these may be reconverted to ASCII code or integers. To avoid noise, filtering is essential (Section 5.3).

Conventionally, a signal amplitude $x(t)$ is plotted versus time t, with $t = 0$ at the beginning of the signal. Usually the biological or electrophysiological signals are analog or continuous, since both x and t are continuous variables. The analog signal $x(t)$ is sampled every T seconds. This amplitude value is denoted by $x(kT)$. When $t = kT$, $x(kT)= x(t)$.

Invariably, the sampling process introduces a quantisation error that is a function of the number of digital bits available to represent the sample [A–D]. Therefore, a quantized sample $\hat{x}(kT)$ is utilised instead of the true sample $x(t)$. T denotes the time step or sample interval and k is the sample number. The notation $x(kT)$ is also denoted by $x(k)$ or x_k, where T is implied.

The *sampling frequency* or *sampling rate* of an analog signal is equal to $1/T$ Hz and should be at least twice the maximum frequency (e.g., for EEG, 256 Hz is sufficient). This is known as the *sampling theorem.* It is necessary to avoid aliasing or overlap. A *sample set* or *data sequence* is a collection of data samples over some range of k. Albeit, generally k lies between $-\infty$ and $+\infty$, practically k is taken between 0 and $(N - 1)$, where N is the total number of data samples.

5.5 Storage and Display

For storage and display, a cathode ray oscilloscope (CRO) may be used. Similar to a pen recorder, a CRO also presents a 2-D graphical display, but the electronic beam can trace repetitive complex patterns at a velocity as high as 10^4 m/s. Any recording that can be done through a polygraph may be displayed through a CRO. The constituent parts are:

i. cathode ray tube (CRT): commonly monochromatic (green or blue) varieties employ one type of phosphor for screen coating
ii. horizontal displacement system
iii. vertical displacement system
iv. power supply system

To plot a given signal against time, a ramp or sawtooth voltage is applied to the horizontal deflection plates and the signal is presented to the vertical deflection plates.

To maintain signal fidelity, sufficient bandwidth and adequate rise time is essential. For lower speeds, bandwidth is more important, while rise time has to be taken care of in the case of faster signals. The product of bandwidth (MHz) and rise time (ns) is about 0.35 and, usually for a good oscilloscope, the product of minimal bandwidth and fastest rise time is 1.7.

Sensitivity alludes to the input required for producing a specific deflection of the CRT spot. Commonly it is expressed in mV/cm. It is inversely related to the bandwidth.

For recording with CRO, a synchronised camera may be used cautiously, or the signal may be concomitantly applied to a pen recorder, or it may be recorded on a magnetic tape for subsequent retrieval and analysis.

CRT storage purveys economic retaining of graphic and alphanumeric display without the need for digital refresh circuitry. Storage CRTs may be bistable phosphor storage tubes, half-tone (variable persistence) storage tubes or a fast transfer tubes.

For isotope scanning, thermography and patient monitoring systems, multicolour CRT with many types of phosphor is used for screen coating.

Multi-channel displays are used commonly, either in the operation theatre or in the ICU. This kind of display can be achieved by using one CRT for each channel, but that becomes too costly. For a 2-channel display, two electron gun CRTs may be used. The cheapest method is the use of an electronic switching system. Another method is a light modulation (instead of a 2-D deflection) system with high frequency sampling.

Non-fade display systems are essential for low frequency waveforms which are common in most biomedical signals. This is obtained by digital memories, whereby a rolling waveform, continuous, bright and flicker-free display is visualised. These systems will require a moderately high sampling rate (e.g., for EKG 500 Hz may be occasionally necessary, but usually a 180 to 250 Hz sampling frequency is enough). For accuracy in signal conversion, the word length should be optimal. Commonly an 8-bit or 1-byte word length is used, so that 2^8 or 256 different levels may be resolved. The memory capacity required will depend on the duration, screen size, rate of display, and sampling frequency (e.g., for respectively 10s, 13 cm, 25 mm/s and 200 Hz specifications, the memory needed will be 2000 word).

5.6 Pattern Classifiers

After storing and displaying signals or images, they often have to be classified according to need or urgency. For instance, in an intensive care unit (ICU), the monitors are usually semi-automated. Most of the time, a patient in the ICU may be in a relatively stable condition. At that point of time no physician may be physically present. In such a case, if the monitor can send an alarm if an abnormal signal (e.g., EKG tracing) is observed, supporting paramedical personnel can immediately call for the concerned specialist doctor. The essential requirement is to have an in-built pattern classifier in the system itself. In any case, e.g., in the case of EEG, automated quantified and categorised data are presented to the user (clinician).

There are various methods of classifying patterns in images or signals. This section purveys only a brief outline of some standard pattern classifiers.

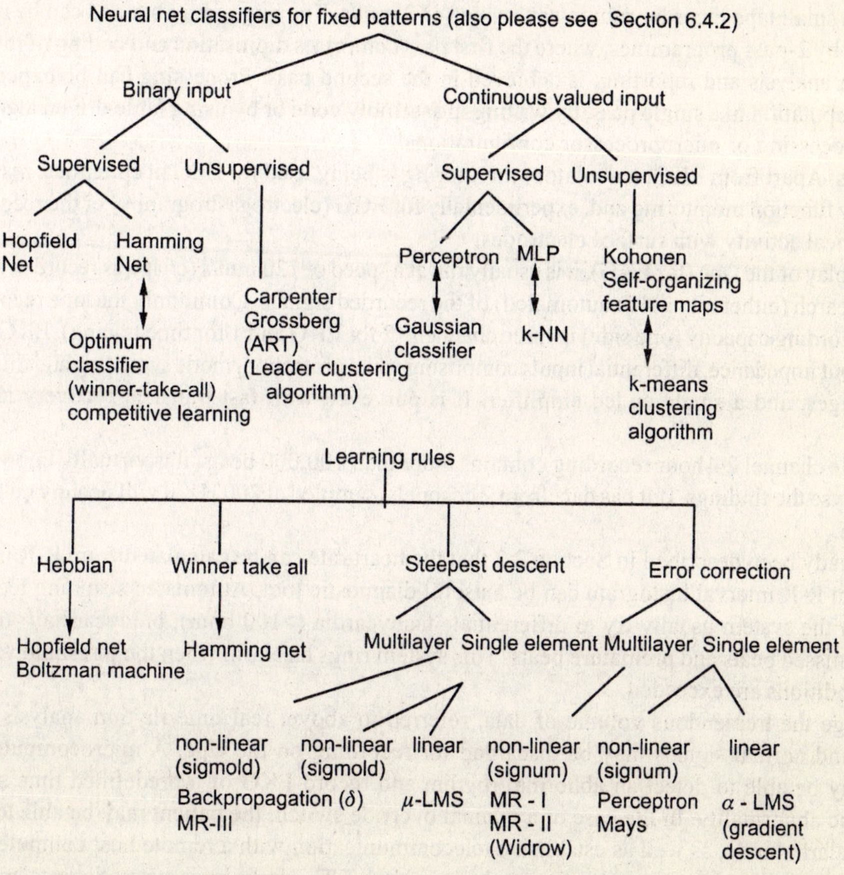

The *signal distance* or *Hamming distance* is the number of corresponding bit positions which differ in two binary words of the same length, e.g., in the two 6-bit words (100101 and 001100) the Hamming distance is 3. If an XOR operation is performed between these, the new word will correspond to the signal distance; therefore, an XOR is often called the distance operation.

The common classifiers for any pattern (signal or image) are:

i. parametric Bayesian classifiers which are used when *a priori* probabilities of occurrences are known
ii. hyperplane classifiers like MLP which use sigmoid (monotonically increasing S-shaped function) functions like hyperbolic tangent
iii. kernel or receptive field classifiers like radial basis function (RBF) which use a kernel function
iv. exemplar classifiers like k-nearest neighbor (k-NN), learning vector quantizer (LVQ) or adaptive resonance theory (ART) which use Euclidean norm

5.7 Ambulatory Monitors

Besides the stationary monitoring systems we have discussed, we also have various ambulatory monitoring systems (e.g., the 24-hour Holter monitoring of EKG). Essentially, the Holter monitor is a multichannel small tape recorder running at a speed of 2 mm/s. The huge data obtained can be replayed and analysed by 2-pass programmes, where the first pass comprises digitisation and coding of raw EKG, while pattern analysis and reporting, is achieved in the second pass. Processing can be expedited by allowing computation in a single pass, by writing in assembly code or by using table driven algorithms, analog preprocessing or microprocessor configurations.

Other uses: Apart from EKG, ambulatory monitoring is being used for EEG in epileptics, respiratory or pulmonary function monitoring and, experimentally for EGG (electrogastrography) or the recording of gastric electrical activity with surface electrodes.

During replay of the tape (for EKG), it is usually run at a speed of 120 mm/s (*cf.* it was recorded at 2 mm/s) for rapid search (either manual or automated) of the recorded signals. Commonly the tape recorder has a 24-hour recording capacity (one side) in three channels (2 for EKG and 1 for time tracing). EKG circuits have high input impedance, differential inputs comprising spike protection, mode switching and differential amplifier stages, and a single ended amplifier. It is purveyed with fast transient recovery and reset options.

As a single channel 24-hour recording contains more than 100,000 beats, it is virtually impossible to visually analyse the findings. If it has data from 2 channels, sampled at 200 Hz, it will occupy no less than 34 Mb space.

It has already been described in Section 2.2 that the heart rate can be calculated from R-R intervals. Therefore, an R-R interval histogram can be a useful diagnostic tool. Automated scanning techniques employed in the system usually try to differentiate tachycardia (>100 bpm), bradycardia (<60 bpm), dropped or missed beats and premature beats. This system rings an alarm when the threshold values set for these conditions are exceeded.

To manage the tremendous volume of data, referred to above, real time rhythm analysis may be performed and normal signals may be discarded for recording on the tape. A microcomputer based program may be able to detect an abnormal rhythm and record EKG of a predefined time span just preceding the abnormality. In the case of a manual override switch, the patient may be able to put the device into alarm mode, as well as establish a telecommunication with a remote host computer (in the health care centre where the patient goes for check up regularly). This is the basis of practicing telemedicine, described in Section 3.3.

5.8 Further Reading

[1] Gonzalez, R.C and Woods, R.E. *Digital Image Processing.* Reading, MA: Addison-Wesley. 1992.

[2] Oppenheim, A.V and Schafer, R.W. *Digital Signal Processing.* New Delhi: Prentice-Hall of India Pvt. Ltd. 1989.

[3] Stearns, S.D and David, R.A. *Signal Processing Algorithms.* Englewood Cliffs, NJ: Prentice-Hall Inc. 1988.

[4] Jay, F. *IEEE Standard Dictionary of Electrical and Electronics Terms*. New York: IEEE Inc., John Wiley & Sons. 1977.

[5] Tretter, S.A. *Introduction to Discrete-Time Signal Processing.* New York: John Wiley. 1976.

5.9 Questions

1. Descriptive Questions

1. What are the various types of recorders available for recording biomedical signals?
2. What are the different types of storage and display systems available?

2. Write Short Notes on

1. Digital recorder
2. Filters
3. CRT
4. Ambulatory monitors

3. Multiple Choice Questions:

Choose the single best response.

1. Usually, biological or electrophysiological signals are analog or continuous, since
   (A) both x and t are continuous variables.
   (B) x is a continuous variable.
   (C) t is a continuous variable.
   (D) none of the above

2. Holter monitor is used for ambulatory
   (A) EEG.
   (B) EMG.
   (C) EKG.
   (D) EOG.
3. A digital filter may be
   (A) bandpass
   (B) bandstop
   (C) infinite impulse response
   (D) any of the above

4. Bayesian classifiers are
   (A) hyperplane
   (B) parametric
   (C) exemplar
   (D) kernel
5. Magnetic tapes may be used for recording from
   (A) CRO
   (B) Holter monitor
   (C) EEG machine
   (D) all of the above

Answers

1. A; 2. C; 3. D; 4. B; 5. D

6

Some Signal Processing Tools

6.1 Introduction

In the last chapter, we learnt how electrophysiological (digital) signals were acquired. In this chapter we will deal with some signal processing tools.

To learn the fundamentals of electrophysiological signal analysis and processing, we consider mostly EEG (which was described in Section 2.3). Electroencephalograms reflect the electrical activity of the brain. Typical EEG waveforms, recorded digitally, during wakeful, slow wave or deep sleep (SWS) and rapid eye movement or paradoxical or dream-sleep (REMS) states are depicted in Fig. 6.1.

There are two contradictory conjectures explaining the generation of EEG signals. The most well-known conjecture is that the EEG is a very complex-looking, virtually unpredictable signal; therefore, it must be produced by a very complex system. The statistical pattern recognition for complex system analysis has been based on the extraction of quantitative features from EEG intervals. These features are supposed to be representative of the test segment and usually include parameters related to the frequency content of the signal.

The second conjecture is that the EEG is a very complex-looking, virtually unpredictable signal; therefore, it must be produced by a relatively simple, non-linear system. EEG signals are assumed to be the output of a deterministic system of relatively simple complexity but one which contains non-linearities. In many models of neural networks processing non-linear wave pulse functions, the non-linear characteristic is approximated by a linearised version, especially when the input signals have a small range of amplitude. However, such oversimplification may ignore essential non-linear properties of the system, like, the generation of higher harmonics. In reality, these networks can be described by a number of coupled non-linear differential equations as a function of time and space. These sets of equations may purvey knowledge of the genesis of any EEG activity. Whether the system will find itself in any stable state depends on the system's parameters and input conditions.

Complex body rhythms are ubiquitous in living organisms. These rhythms arise from stochastic, non-linear and biological mechanisms interacting with a fluctuating environment. Disease often leads to the alteration of a normal rhythm to a pathological rhythm. Mathematical and physical techniques combined with physiological and medical studies are increasing our understanding of the rhythms of life. It has been found that chaotic dynamics might be easier for the body to control than stochastic dynamics.

The term 'dynamical disease' captures the notion that abnormal rhythms, which could be either more irregular or more regular than normal, arise owing to the modifications in physiological control systems that lead to bifurcations in the dynamics. What is more important in distinguishing health from disease is that there is a change in the dynamics from what is normal, rather than the regularity or irregularity of dynamics.

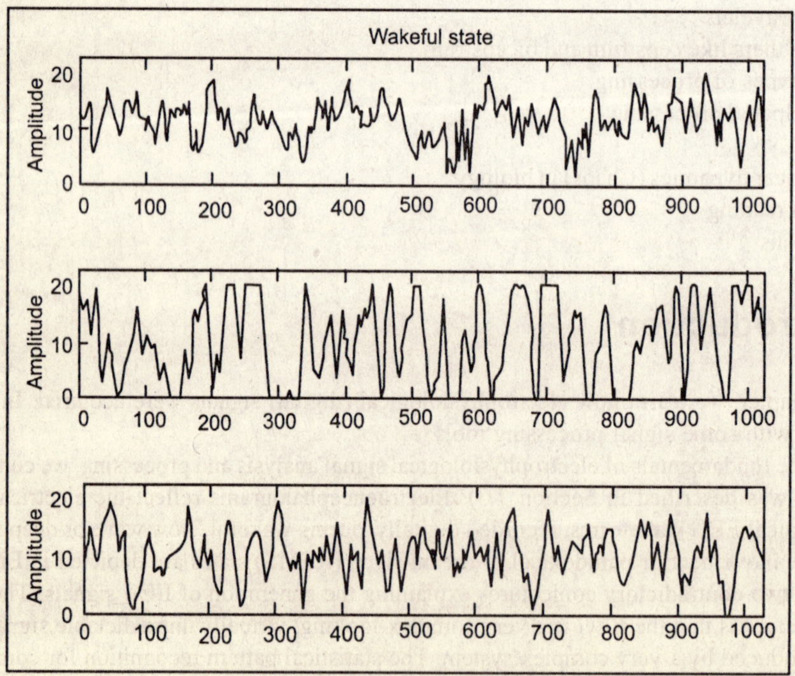

Fig. 6.1 Typical digital EEG signals in wakeful, deep sleep and REM sleep stages

Chapter 2 dealt with various biopotentials and EEG of which Fig.6.1 is an illustrative example. EEG reflects some (but not all) aspects of the brain activity, whereas for understanding the other aspects of the brain other techniques like CT/MRI scans are essential. Consult Chapter 18 for a comparison.

6.2 Electrophysiological Signal Processing

As mentioned in Chapter 5, electrophysiological signals need to be analysed, unwanted interference or noise eliminated or reduced, and transformed into a more informative form, before it can be productively used. Some processing methodologies are described in Chapters 11 and 18.

6.2.1 Electrophysiological Signal Analysis

Biological signal analysis may be discussed under the heads of temporal (parametric and non-parametric), spatial (topographic), artificial neural network (ANN)-aided and non-linear processing methods. These are briefly described in the subsequent sections. However, for most bioelectrical signals like EEG, analysis comprises not just quantification, but also feature extraction and classification. This may involve various pattern recognition techniques as well.

6.2.2 Noise Reduction

Generally, noise reduction is achieved by using filters (both analog and digital, see Section 7.5 and Section 5.3). However, taking appropriate precautions during recording an event of biopotential generation is absolutely necessary for getting proper (interference-free) information.

6.2.3 Making Signals more Informative

For getting additional information from a given recording of an electrophysiological signal, it is important to understand the nature of the signal.

Spatiotemporal contextual information can be incorporated for automated EEG analysis through syntactic analysis techniques. Here, EEG is represented as a series of elementary patterns called *tokens*. After sub-dividing an EEG tracing into 1-second intervals, a sequence of unique characters may be called a *label*. The resulting sequence of labels is usually known as a *sentence* which is parsed by a definite grammar. A *grammar* consists of a set of rules that governs the merging of tokens into higher-order recognisable entities in the input data.

A severe drawback in this approach lies in the amount of heuristics involved in writing the grammar. Also, it is almost impossible for an untrained electroencephalographer (EEGer) to inspect the grammar and make the necessary modifications. Knowledge-based approaches (Sections 19.4.1 and 19.4.2) purvey a way of increasing the involvement of the expert (EEGer) in the design process and also allow the utilisation of contextual information to a larger degree. These systems (expert systems or ESs) apply a body of knowledge (knowledge-base) to the input data and subsequently derived facts (data-base). Typically, the knowledge base comprises IF<premise>THEN<action> type rules which presumably reflect the human expertise. Another advantage of this type of flexible problem-solving approach is that the collection of rules may be modified from time to time. Knowledge-based enhancement of EEG signals and parallel processing are employed to save time and effort, as well as to increase the accuracy of interpretations.

Even today, textbooks of medicine, neurology and psychiatry do not attach as much importance to EEG as they do to an ECG. The EEG is considered to be quite non-specific, except in a few cases of epileptic disorders. Moreover the same EEG recording elicits different interpretations not only from different experts (inter-observer variations), but also from the same expert (intra-observer variations) at different times. On the other hand, journals, specialised textbooks and reference books on EEG stress the advantages of the various new computer-aided (quantitative or qEEG) methods for analysis. As yet there is no consensus on the appropriate mathematical analysis to be used. Each method has its inherent advantages and disadvantages. A neuropsychiatrist is not very likely to afford the time to learn all the techniques. In the near future, this may lead to laboratory-dependent qEEG diagnosis. Initially what was started with the aim of simplifying and standardising EEG interpretations, have, with time, grown into a much more complex and ever-expanding exercise. However, without these computations, we could not have learnt about the finer details of the EEG and brain functions.

6.3 Temporal Processing

6.3.1 Non-parametric Methods

EEG analysis may be done, in the temporal (time) domain, non-parametrically by amplitude distribution, interval analysis, interval–amplitude analysis, correlation analysis, power spectra, cross-spectral analysis, bi-spectral analysis, and Walsh and Hjorth slope descriptors.

Amplitude analysis

A biological signal like EEG may be characterised by the distribution of its amplitude and its moments. In amplitude distribution of EEG, the first query is whether it is normal or Gaussian (Section 20.2.1). A common test for Gaussianity is the χ^2-test (Section 20.2.4). However, for small EEG samples, the Kolmogorov–Smirnov test is more informative than the χ^2-test. The prerequisites for the application of these tests are stationarity and independence of adjacent samples. However, a sampling frequency (Section 5.4) of about 20 Hz is necessary to achieve even an acceptably small degree of (auto)correlation between adjacent samples.

Gaussianity may also be checked by measuring skewness and kurtosis, after taking into account the possibility that adjacent samples may have a large (> 0.5) degree of correlation. These have to be checked from the available statistical tables. Measuring the surface of rectified EEG waves may also give some information of its amplitude content, but is not used routinely.

Interval analysis

Interval (period) analysis of EEG waveforms has been used to study its statistical properties. It has also been used along with other methods like autocorrelation and power spectra. Its main purpose has been to monitor long-term EEG changes in anaesthesia, actions of psychoactive drugs and in sleep research. Here, the zero-crossings of the original EEG, as well as those of the signal's first and second derivatives are computed. One major drawback of this method is its sensitivity to high frequency noise in the estimations of zero-crossings, while another shortcoming is the underestimation of low frequency and overestimation of high frequency components. These problems may be overcome by computing the zero-crossing intervals only within predefined frequency bands. On the other hand, the chief advantage is its ease of computation.

Interval–amplitude analysis

In interval–amplitude analysis, EEG is decomposed into waves or half-waves, defined both in time (by the interval between zero-crossings), and in amplitude (by the peak-to-trough values). It is also known as *sequential analysis* since the amplitude and interval duration of successive half-waves are analysed, displayed and stored in sequence in real time. This requires a sampling frequency above 250 Hz.

The amplitude and the interval duration of a half-wave are defined by the peak–trough differences in amplitude and time, while those for a wave are measured by the mean amplitude and the sum of the interval durations of two consecutive half-waves.

Correlation analysis will be discussed under autoregressive moving average.

Power (density) spectra or fast Fourier transform (FFT)

Nearly two hundred years ago, Joseph Fourier (a contemporary of Galvani, Volta, Euler and Bernoullis) defined the Fourier series as an expansion of a periodic function into sinusoid ({i.e., sine and cosine waves) functions according to a specific rule. The Fourier series is a special case of Fourier transform.

The EEG signal can be considered as a function as it is composed of numerous waveforms. By the Fourier transform, the original signal (amplitude as a function of time) is converted to an amplitude spectrum – a graph of amplitude as a function of frequency. Usually the amplitude values are squared to give rise to power (density) spectra. It may be a continuous spectrum in which the graph is a continuous line over a range of frequency or a discrete spectrum (sometimes known as a line spectrum) in which only specific frequencies are present and their amplitudes are represented by vertical lines at those points on the frequency axis. This is the *discrete Fourier transform (DFT)* of the signal. When it is computed in windows (e.g., epochs of 4 seconds), it is also called *windowed DFT* or *WDFT*.

The FFT coefficients (power spectra) of the EEG waveforms given in Fig. 6.1 are presented in Fig. 6.2.

Cooley and Tukey presented an algorithm for the computation of a DFT applicable when N is a composite number (product of two or more integers) and called it a *fast Fourier transform (FFT)* algorithm. Presently an FT can be done with the help of a personal computer using the FFT algorithms. The advantages of these spectra are its transparency and reversibility (by inverse DFT, the original signal can be regained). A sinusoidal function $X(t)$ is completely described by three parameters – amplitude A, frequency f and the phase f as $X(t) = A \sin(2ft + f)$, where t is the time. Spectral analysis is a method of calculating the amplitude A_n and phase f_n corresponding to each frequency component f_n that is contained in the EEG epoch. The power spectrum of the EEG is a measure of the relative magnitudes of the A_n^2s, not taking into account the phase information. In the case of a complete direct evaluation of an N-point DFT, $4N^2$ real multiplications and $(4N^2 - 2N)$ real additions are required. For the direct computation of the DFT, the amount of computation, and thus the computation time, is approximately proportional to N^2 and can become very great for large values of N. The basic principle is decomposing the computation of the DFT of a sequence of length N into successively smaller DFTs. It may be either decimation-in-time, where the sequence $X(t)$ is decomposed into successively smaller sub-sequences or decimation-in-frequency, where the DFT coefficient $X(k)$ is decomposed into smaller sub-sequences. The decimation-in-time algorithm is best applicable when N is an integral power of 2 i.e., $N = 2^r$.

FFTs are not superior to time-domain transforms; rather they are more informative than the conventional analog recordings which are still preferred by clinicians. Any transform, being derived from the original signal, is always likely to lose some information. However, frequency domain processing makes the findings more objective (quantified) and hence, observer-independent. Power density spectra measurement is a form of quantitative EEG (qEEG), albeit not the only one.

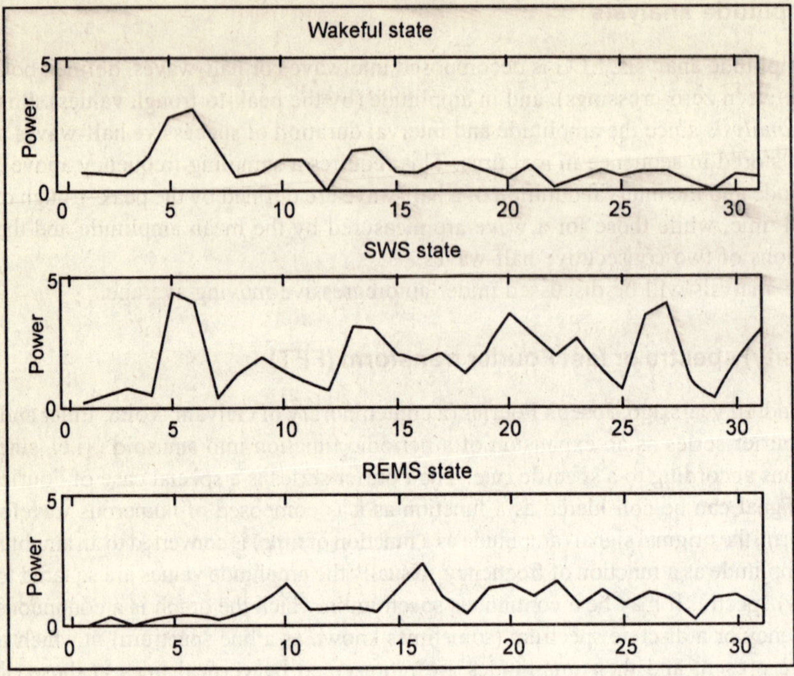

Fig. 6.2 FFT tracings of the EEG signal in Fig. 6.1

Cross-spectral analysis

The quantification of the relationship between different EEG signals may be achieved by cross-spectral analysis. The cross-power spectrum $C_{xy}(f)$ is the product of a smoothed DFT of one signal and the complex conjugate of the other. This complex quantity has both magnitude and phase and is represented by:

$$C_{xy}(f) = C_{xy}(f) \cdot \exp\left[j\phi_{xy}(f)\right]$$

where $j = \sqrt{-1}$. The function of frequency $\phi_{xy}(f)$ is the phase spectrum. A normalised quantity coherence function is denoted as:

$$\mathrm{Coh}_{xy}(f) = \frac{\left|C_{xy}(f)\right|^2}{P_{xx}(f)P_{yy}(f)}$$

The counterpart of coherence is the phase function. Here *phase* is a mathematical idea alluding to the proportion of the period of sine wave component of a signal as obtained by Fourier analysis.

Bi-spectral analysis

Another tool, for the analysis of signals like EEG, the bi-spectrum, is the Fourier transform (FT) of the third-order cumulant (TOC) sequence, generally used (a) to extract information from the signal pertaining to deviations from Gaussianity and (b) to detect the presence of non-linear properties and quadratic phase coupling. Since TOC and bi-spectrum are both zero for a stationary Gaussian process, bi-spectrum purveys a natural measure of Gaussianity. Also, a bi-spectrum can detect phase coupling (i.e., the presence of quadratic non-linearity) and the degree of phase coupling can be quantified using the bi-coherence index, i.e., a normalised bi-spectrum. *Bi-spectrum* or *bi-spectral density* may also be defined as the following 2-D transform (FT_2) of the autocorrelation function $R_{xx}(t_1,t_2)$:

$$B_{xx}(f_1,f_2) = FT_2[R_{xx}(\tau_1,\tau_2)]$$

Some rare EEG rhythms (e.g., μ waves) have significant relations between harmonics of 10 Hz (at 5, 20 and 30 Hz) and the psychomotor variant having relations between 6, 12, 18 and 24 Hz. Occasionally, a rhythms (Berger waves, see Section 2.3) may also show a significant bi-spectrum. Bi-spectra have been computed to detect phase coupling in the cortical and hippocampal EEG of a rat during various vigilance states. For EEG recordings from the hippocampus, significant phase coupling was obtained during REM sleep between the frequency components 6–8 Hz associated with the q rhythm. During slow wave sleep, EEG from the frontal cortex or hippocampus exhibits a larger deviation from Gaussian distribution than those during quiet waking and REM sleep. So, bi-spectral analysis yields extra information not obtainable from power spectrum.

Walsh and Haar transforms are not-so-popular methods of computing power spectra and will not be discussed here.

Slope analysis

The moments of spectral density $S_{xx}(f)$, may be used in the application of Hjorth descriptors. Hjorth (1975) has developed hardware for real time computation of spectral moments. Since the *n*th-order spectral moment may be defined as

$$a_n = \int_{-\infty}^{\infty} (2\pi f)^n S_{xx}(f)df,$$

zero-order moment, is also termed as

$$a_0 = \int_{-\infty}^{\infty} S_{xx}(f)df$$

(this is equal to the variance σ^2, please see Section 20.2.5), second-order moment, is represented by , and the fourth-order moment is denoted as . Here, the spectral moments (statistical properties) are allowed to vary in time, so, this form of analysis can be applied to non-stationary signals. The descriptors are derived parameters like activity, denoted by A, ($A = a_0$), mobility, represented by M, ($M = [(a_2/a_0)]^{1/2}$), and complexity, symbolised by C, ($C = [(a_4/a_2) - (a_2/a_0)]^{1/2}$). Just as a_0 is equal to the variance s^2 of a signal like EEG, a_2 is the variance of the signal's first derivative, and a_4 is the variance of the signal's second derivative. However, despite relative ease in computing these descriptors, they can give a valid description of EEG only when the signals have a symmetric probability distribution function with only one maximum.

6.3.2 Parametric Methods

EEG signals may also be analysed, in the time domain, parametrically by the computation of power spectra through autoregressive or ARMA modelling, inverse autoregressive filtering, time-varying signals – Kalman filtering, segment analysis, mimetic analysis, matched filtering or template matching, time–frequency analysis through wavelets or convolution.

Autoregressive moving average (ARMA)

Correlation analysis (more time-consuming and expensive) has been a forerunner of FFT (non-parametric). However, autocorrelation (compared to cross-correlation) function analysis has been revived due to parametric analysis of autoregressive models.

These methods are known as parametric since the EEG signals used are described by a mathematical model denoted by a set of parameters. A non-specific and inessential link may exist between the mathematical model and the biophysical model for generation of a particular waveform (e.g., a waves between 8 and 12 Hz). In this method, apart from spectral computations, detection of transient non-stationarities like epileptic spikes or sharp waves and sub-division into quasi-stationary segments are also possible. Also, in these techniques, EEG data may be substantially reduced. An autoregressive model can describe an EEG signal with few coefficients. By following the magnitudes of these coefficients, it is possible to trace the time-varying properties of the signal. Further, these coefficients can be used for cluster analysis (classification of EEG power spectra).

There are two ways of representing the signal. In the continuous case, the signal *X(t)* is assumed to originate by filtering through a transfer function *H(s)* on a noise source with a flat spectrum within the frequency range in question. In the discrete case, the EEG signal is presented as a set of samples *X(k)*, obtained by filtering on an input noise signal *E(k)* with mean zero.

The filter in the autoregressive moving average (ARMA) model is expressed by a linear differential equation

$$a_0 x(k) + a_1 x(k-1) + \cdots + a_p x(k-p) = b_0 e(k) + b_1 e(k-1) + \cdots + b_q e(k-q)$$

where $q \# p$. The expressions *x(k)* and *e(k)* are related by the sets of coefficients $a_1 \ldots a_p$ and $b_1 \ldots b_q$, with $a_0 = 1$. If $b_i = 0$ for $i = 1 \ldots q$, then we get the AR model:

$$x(k) + a_1 x(k-1) + \cdots + a_p x(k-p) = e(k)$$

So, here the problem is that of computing the coefficients. However, the minimum number of coefficients to be computed must be known beforehand.

The power spectra may be computed rather easily with the AR or ARMA model. By spectral parameter analysis (SPA), an estimation of the model parameters, as well as the best spectral depiction can be achieved. In the ARMA model, spectral density of a signal sampled with sampling interval is:

$$P_{xx}(f) = \sigma^2 \left| \frac{\sum_{i=0}^{q} b_i (\exp(-j2\pi f i \Delta t))}{\sum_{i=0}^{p} a_i (\exp(-j2\pi f i \Delta t))} \right|^2$$

In the case of an AR model, the spectral density is:

$$P_{xx}(f) = \frac{\sigma^2}{\left|\sum_{i=0}^{p} a_i (\exp(-j2\pi f i \Delta t))\right|^2}$$

Inverse autoregressive filtering

In the inverted form, an AR model can render the inverse autoregressive filtering operation. Assuming the origin of EEG from a stationary process, it is possible to approximate it as a filtered noise with a normal distribution. Consequently, passing such an EEG signal through the inverse of its estimated autoregressive filter should result in a normally distributed noise *N* with mean zero and variance σ^2. The null hypothesis states that an EEG signal can be assumed to be stationary and described in terms of the properties of the estimated noise $\hat{e}(k)$, resulting from inverse autoregressive filtering:

$$\hat{e}(k) = x_k + \sum_{i=1}^{P} a_i x(k-i)$$

The EEG signal is accepted as non-stationary for $t = nT$ if the null hypothesis can be rejected (i.e., if $\hat{e}(k)$ deviates at some probability level from a noise with normal distribution). To smoothen the detection, instead of $\hat{e}(k)$, a detection function $d(k)$ is used, where $d(k) = \sum_{n=k-m}^{k+m} \left[\frac{\ddot{e}(n)}{\hat{\sigma}}\right]^2$. As the square of a normally distributed variable (with variance 1) follows a χ^2 distribution, $d(k)$ should also have a χ^2 distribution with $(2m+1)$ degrees of freedom (Section 20.2.4).

Kalman filtering

For analysing time-varying signals, the Kalman estimation method of tracking may be applied. The input signal to a hypothetical processor responsible for generating the EEG signal is assumed to be a normally distributed noise *e(k)*. An autoregressive model is assumed to express the observed EEG signal. To reach the estimates of the model coefficients, using formerly measured data, a recursive algorithm known as the *Kalman filter* is applied. Here updating is done with new samples of the time series. However, unless the order of the model and initial conditions are chosen properly, implementation will not be easy. One of the uses of this technique has been for segmental analysis of EEG signal, in terms of unvarying statistical properties. Autocorrelation based adaptive segmentation of EEG signals have also been tried.

Mimetic analysis

If automated EEG analysis mimics routine visual analysis, as done by EEGers, it is known as *mimetic analysis*. This is another (syntactic) way of interval–amplitude analysis where half-waves and minimal descriptors correspond to linguistic characters or letters, significant series like sleep spindles match words, segments comprising wave series like rhythms relate to paragraphs and ensembles of segments tally with sections or chapters. Unlike spectral analysis, it combines feature extraction with segmentation and logical categorisation. Here, the extracted features lead to construction of tables and graphs depicting the synoptic distribution of various features in an EEG epoch, followed by the estimation of statistical properties for many epochs and derivations.

Matched filtering or template matching

Matched filtering or *template matching* is a mode of pattern analysis where a particular pattern or template in the EEG signal is detected by cross-correlation between the signal *X(t)* and an *a priori* defined pattern *M(t)*. The estimator reaches the maximal value (1) when *M(t)* = *X(t)*, and both signals are aligned perfectly along the time axis.

Wavelets

Since the EEG is stationary only for relatively brief periods, time-varying spectral analysis (segmentation principles applied to spectral arrays) has been used. Subsequently, optimal analysis in time and frequency domains simultaneously has also been tried. As the phase information cannot be obtained in FFT, wavelets are often used (also for image and signal compression). First, a series of curves is generated, each of which has a width corresponding to a particular EEG frequency. These synthetic curves are then convolved (superimposed and multiplied while stepping sequentially through the signal). Each time the wavelet template matches the wave component in the signal of interest, it produces a large product function that can allow a sequential identification of the frequency content in a rapidly changing signal.

Wavelets typically have a waveform of damped oscillation where $\int_{-\infty}^{\infty} \psi(t)dt = 0$ and,

$\psi_{s,u}(t) = \frac{1}{\sqrt{s}}\psi\frac{(t-u)}{s}$ s being the scale and u the translation variable. From this definition, a set of orthogonal wavelets can be constructed that forms an orthonormal basis and a given EEG signal may be denoted by the corresponding wavelet coefficients. Since the set of wavelets is limited, for obtaining a large set of basis functions, scaling, translating and modulating a window function *g(t)* is necessary. By definition,

$g(t) = \frac{1}{\sqrt{s}} g\frac{(t-u)}{s} e^{i}$, where in the time domain, *g(t)* is concentrated around *u* with a width proportional to *s*; and in the frequency domain, *g(t)* has its energy concentrated around ξ and spread proportional to 1/*s*. The minimal time–frequency variance corresponds to a Gaussian *g(t)*. Practically, the EEG being a real discrete time series, the basis function is usually expressed as: $g_{(\gamma,\phi)}(n) = K_{(\gamma,\phi)} g_j(n-p)\cos(2\pi\frac{k}{N}n + \phi)$,

where the index = *(j,k,p)* is the discrete analog of (ξ, s, u) of the preceding equation, $N = 2^L$ is the number of samples and *p* and *k* parameters are sampled at intervals of 2^j. This is an iterative procedure.

Others types of analyses

Quantification has also been done with cepstrum (power spectrum of the logarithmic power spectrum) and bicepstrum analyses of EEG.

Apart from these, an approach to visual evaluation of long-term EEG recordings has been based on multi-channel adaptive segmentation, subsequent feature extraction, and automatic classification of acquired segments by fuzzy cluster analysis (fuzzy c-means algorithm) and distinguishing of the segments so identified by colour directly in the EEG record.

6.4 Other Types of Processing

6.4.1 Spatial Processing

Spatial analysis is used for topographic display. Duffy (1994) discusses the role of quantified EEG or quantified neurophysiology (qNP) or neurometrics or brain electrical activity mapping (BEAM) which gives colour coded topographical representation of the FFT values in different regions of the brain. The disadvantages of it are that controlling the artefacts are quite difficult, huge data sets have to be managed, classic EEG and multivariate statistics are essential for proper inferences.

Another approach to measure the scalp potentials due to dipole sources has been finite-element modelling (Yan et al. 1991) of the human head.

Before going into brain mapping, we have to understand the genesis of EEG. We have already learnt about neuromuscular modelling in Section 4.4. Synchronously occurring post-synaptic potentials generate scalp EEG. Other membrane potentials like action potentials (Section 1.2.2) do not contribute to EEG signals because of the following reasons.

(i) Membrane potential variation caused by an action potential generates a field which is equivalent to that of a single dipole perpendicular to the membrane because the piece of membrane that is depolarised at any instant of time is very small. In contrast, that of an electrotonically conducted post-synaptic potential extends, at any instant, over a larger portion of the membrane; thus generating a field corresponding to that of a dipole layer with dipoles perpendicular to the membrane surface. The latter attenuates with distance less rapidly than the former.

(ii) Action potentials, due to their short duration (1–2 ms), tend to overlap much less than do post-synaptic potentials (excitatory and inhibitory) which last longer (about 10–250 ms). When action potentials do occur simultaneously, e.g., a short stimulus exciting a group of fibres), the field of the action potential may also be recorded at relatively large distances as a compound action potential. The post-synaptic potentials contribute primarily to the generation of extra-cellular field potentials in the cerebral cortex.

The main factor causing synchrony of the potentials is the structural nature of the nervous tissue (a combination of interlocked excitatory and inhibitory populations). The neural masses act as a cortical macroscopic source, behaving like a dipole layer, directed perpendicular to the cortical surface. The dynamic properties of the rhythms (like frequency) depend upon the parameters of neuronal population viz., feedback gains, strength of connections, time properties of synaptic potentials and non-linear properties i.e., thresholds of saturation). The propagation of electrical activity in the cortex from small 'epicentres' may cause time dilatation i.e., an EEG transient recorded at the cortex will seem to be of shorter duration than the transient recorded simultaneously at the scalp, as in epileptiform spikes. This happens as a moving dipole seen from a distance can be observed over a longer time than when seen from nearby. The surface of the cerebral cortex is composed almost entirely of a mat of dendrites from neurons in the lower layers of the cortex. Dendrites may be conducting processes, but they are usually the sites of non-propagated hypopolarising and hyperpolarising local potential changes. When the excitatory endings become active, current, in the rest of the dendritic mass and the neuronal cell body (soma), flows from the current-sources into the current-sinks. The reverse happens in case of an inhibitory impulse. The cell–dendrite relationship is that of a constantly fluctuating dipole. Current flow in this dipole would be expected to produce waveform potential changes in the volume conductor. When the sum of the dendritic activity is negative compared to the cell, the cell is hypopolarised and hyperexcitable; when it is positive, the cell is hyperpolarised and hypoexcitable. The cerebellar cortex and the hippocampus also produce characteristic fluctuations in surface potentials because of the complex, parallel dendritic processes located subpially over a layer of cells there. The brain wave intensity is closely related to the degree of activity in the thalamic or brainstem portion of the ascending reticular activating system.

The 'inverse problem' which is to determine the intracerebral sources, given a measured potential distribution at the scalp, has no unique solution. According to the modern approach, the intracerebral source is assumed to be an equivalent current dipole localised within the brain. To obtain possibly significant results, it is mandatory to use a model with inhomogeneities (i.e., with shells of different conductivities) to account for the layers surrounding the brain, viz., the cerebrospinal fluid (CSF), the skull and the scalp.

The EEG is a complex signal and its statistical properties depend on both time and space. However, the ever-changing EEG signals can be analytically subdivided into representative epochs. Topographical consideration of the montages is essential for deriving useful information from the EEG.

6.4.2 ANN (Artificial Neural Network)

Before defining ANN, let us see what a biological (natural) neuron is. This will be followed by a few more related definitions.

Biological neuron

A neuron is the structural unit of the biological nervous system, which processes all information, while the synapse is its functional unit. Artificial models of neurons contain not only the neural cell body but also the input synapses and a mechanism for training them.

Artificial neural (or neuronal) networks (ANNs)

ANNs are massively parallel interconnections of simple processing elements that interact with objects of the real world in a manner similar to biological systems. All information is encoded among the connection weights in a distributed manner. They are non-parametric, adaptive, robust, fast and optimal. Generally ANNs consider a fixed topology of neurons connected by links in a predefined manner. Knowledge-based networks constitute a special class of ANNs that consider crude domain knowledge to generate the initial network architecture, which is later refined in the presence of training data. This process reduces searching space and time while the network traces an optimal solution.

Neural processing

In the human brain, about 10^{10} to 10^{11} neurons, with their 10^{14} to 10^{16} interconnections, undertake parallel processing. Neurons are failure-prone devices with redundancy. Approximate matching is done very well because biological knowledge is represented in a diffuse and fluid way.

Learning

Formally, a device learns a concept if it can, given positive and negative examples, produce an algorithm that will classify future examples correctly with a probability of $1/h$. The complexity of learning a concept is a function of 3 factors: error tolerance (h), the number of binary features present in the examples (t) and the size of the rule necessary to make discrimination (*f*). If the number of training examples required is polynomial in *h, t, f*, then the concept is said to be learnable.

Supervised

In supervised training, both input and expected output is supplied at training.

Unsupervised

Only input samples are given in unsupervised training, e.g. the winner-take-all or mutual inhibition process.

Combined

Combined learning is a mixture of supervised and graded learning.

Adaptation

Adaptation is a changing learning process through reward and punishment. In a genetic algorithm (GA), learning occurs through a selection process beginning with a large population of random programs. Adaptation is achieved here by 'mutation', as in evolution. In many cases, the weights may be frozen to stop further learning, once the training is complete. However, it sometimes is necessary to continue to adapt itself to changing conditions e.g. if an ANN is supposed to control attitude of an orbiting spacecraft whose mass reduces with fuel spent.

After learning the basic definitions pertaining to ANNs, we now have to learn about the components and the mathematical functions governing their actions.

Adaptive coefficients

Adaptive coefficients are values of the previous computations (weights) of a processing element stored in its local memory, which modify subsequent computations.

Connection

A connection is a signal transmission pathway between processing elements, corresponding to the axons and synapses of neurons in the biological brain that connects the processing elements into a network.

Learning law

An equation that modifies all or some of the adaptive coefficients (weights) in a processing element's local memory in response to input signals and the values supplied by the transfer function. The equation enables the network to adapt itself to examples of what it should be doing, to organise information within itself, and thereby learn.

Processing element

A processing element is generally an artificial neuron in a neural network, comprising a small amount of local memory and processing power. The output from a processing element is fanned out and becomes the input to many other elements.

Scheduling function

The scheduling function determines if and how often a processing element is to apply its transfer function.

Transfer function

A mathematical formula which, among other things, determines a processing element's output signal as a function of the most recent input signals and the adaptive coefficients (weights) in local memory. The transfer function includes the learning law of the processing element.

Transformation

Mappings or associations of objects or representations in one set (viz. written words) with objects or representations in another set (viz. spoken sounds) in accordance with some rule, which is typically not known to a human programmer, but is implicit in the training data.

Weight

A weight is an adaptive coefficient within a processing element, associated with a single input connection, depending on the network's design and the information it has learned.

The common transfer functions and their uses are:

i. *Hard limiter (signum)* used in perceptron, madalines. (linear)
ii. *Threshold logic* e.g., heaviside function (special sigmoid function with $\beta \rightarrow \infty$).
iii. *Sigmoid* most commonly used, e.g., in backpropagation. (non-linear)

The most commonly used decision regions are:

i. *Single layer* ½ plane bounded by a hyperplane.
ii. *2-layer* convex open or closed region.
iii. *3-layer* arbitrary (complexity according to number of nodes)

Also see Section18.6 for ANNs as classifiers.
A typical and commonly used ANN (multi-layer perceptron or MLP) is given in Fig. 6.3.

For a single layer perceptron, the algorithm is:

Step 1 Initialise weights and threshold

Set $w_j(0)$ $(0 \le i \le N-1)$ and θ to small random values. Here, $w_j(t)$ is the weight from input i at time t and θ is the threshold in output nodes

Step 2 Present new inputs and desired outputs

Present new continuous valued input $x_1, x_2, \ldots, x_n$ along with desired outputs $d(t)$

Step 3 Calculate actual outputs

$$y(t) = f_n\left(\sum_{i=0}^{N-1} w_i(t)x_i(t) - \theta\right)$$

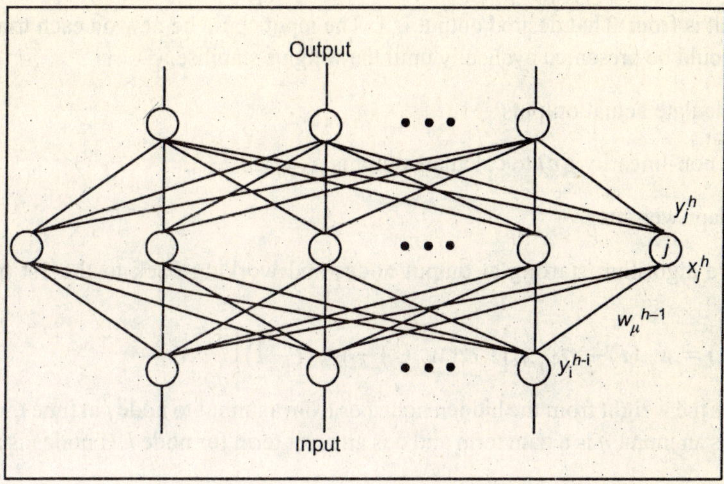

Fig. 6.3 A multi-layer perceptron (MLP)

Step 4 Adapt weights

$$w_i(t-1) = w_i(t) + \eta[d(t) - y(t)]x_i(t), \quad (0 \le i \le N-1)$$

$$\text{dt} = \begin{cases} +1 \text{ if input from class A} \\ -1 \text{ if input from class B} \end{cases}$$

In these equations, η is a positive gain fraction < 1 and *d(t)* is the desired correct output for the current input. Note that weights are unchanged if correct decisions are made by the net.

Step 5 Go to Step 2

The most common algorithm used for a multi-layer perceptron is the backpropagation algorithm, which is briefly described below. It is an iterative gradient training algorithm designed to minimise mean square error (MSE) between the actual output of a multi-layer feed-forward perceptron and the desired output. It requires continuous differentiable non-linearities. Sigmoid logistic non-linearity is used in

$$f(\alpha) = \frac{1}{1 + e^{-(\alpha - \theta)}}$$

Step 1 Initialise weights and offsets

Set all weights and node offsets to small random values.

Step 2 Present new inputs and desired outputs

Present a continuous valued input vector $x_0, x_1, \ldots, x_{N-1}$ and specify desired outputs $d_0, d_1, \ldots, d_{N-1}$. If the net is used as a classifier, then all desired outputs are typically set to zero except for that corresponding to the class the input is from. That desired output is 1. The input could be new on each trial, or samples from a training set could be presented cyclically until the weights stabilise.

Step 3 Calculate actual outputs

Use sigmoid non-linearity *f(a)* to calculate outputs $y_0, y_1, ..., y_{N-1}$.

Step 4 Adapt weights

Use recursive algorithm starting at output nodes and working back to the 1st hidden layer. Adjust weights by

$$w_{ij}(t+1) = w_{ij}(t) + \eta\delta_i x_i' \left[+ \alpha(w_{ij}(t) - w_{ij}(t-1)\right]$$

where $w_{ij}(t)$ is the weight from the hidden node *i* or from an input to node *j* at time *t*. x_i' is either the output of node *i* or is an input. η is a gain term and δ is an error term for node *j*. If node *j* is an output node, then

$$\delta_j' = y_j(1-y_j)(d_j - y_j)$$

where d_j is the desired output of node *j* while y_j is the actual output. If node *j* is an internal hidden node, then

$$\delta_j = x_j^i(1-x_j^i)\sum \delta_k w_{jk}$$

where *k* is the overall nodes in layers above node *j*. Convergence is faster by adding a momentum α $(0 < \alpha < 1)$.

Step 5 Go to Step 2

6.5 Non-linear Dynamics (Chaos) in Biology

As explained earlier, complex body rhythms, ubiquitous in living organisms, arise from stochastic, non-linear and biological mechanisms interacting with a fluctuating environment. Chaotic dynamics might be easier for the body to control than stochastic dynamics.

Electrical stimulation of complex dynamics in cardiac and neural systems using chaos-control techniques has led to the regularisation of complex rhythms.

There is a wide spectrum of dynamic behaviour associated with both normal and pathological physiological functioning. Extremely regular dynamics are often associated with disease, including periodic (Cheyne–Stokes) breathing, certain abnormally rapid heart rhythms, cyclical blood diseases, epilepsy, neurological tics and tremors. However, regular periodicity can also reflect healthy dynamics – for example in the sleep–wake cycle and menstrual rhythms. Finally, irregular rhythms can also reflect disease. Cardiac dysrhythmias such as atrial fibrillation and frequent ectopy, and neurological disorders like post-anoxic myoclonus, are often highly irregular. The term 'dynamical disease' captures the notion that abnormal rhythms, which could be either more irregular or more regular than normal, arise owing to the modifications in physiological control systems that lead to bifurcations in the dynamics. What is more important in distinguishing health from disease is that there is a change in the dynamics from what is normal, rather than regularity or irregularity of dynamics.

Electroencephalograms (EEGs) reflect the electrical activity of the brain. Even when they are analysed from healthy individuals, they manifest chaos in the nervous system. Sarbadhikari and Chakrabarty[2] studied the non-linear/chaotic interactions that place in the brain in order to point towards new theoretical models incorporating the oscillation caused by the same neurotransmitter acting on its different receptor subtypes. EEGs are likely to be produced by a non-linear system, since a non-linear system with at least 3 degrees of freedom (or state variables) may exhibit chaotic behaviour. Furthermore, such systems can have multiple stable states governed by 'chaotic' ('strange') attractors. A key feature of chaotic systems is the presence of an infinite number of unstable periodic fixed points, which are found in spontaneously active neuronal networks (e.g., epilepsy).

The brain has chemicals called neurotransmitters that convey information through the 10^{16} synapses residing there. However, each of these neurotransmitters acts through various receptors and their numerous subtypes, thereby exhibiting complex interactions.

While in epilepsy the role of chaos and EEG findings are well-proven, the role of chaos is also slowly gaining grounds in another condition, i.e., depression. The multifarious roles of exercise, neurotransmitters and (cerebral) hemispheric lateralisation, in the case of depression, are also being established. The common point of reference could be non-linear dynamics.

An interesting point to remember is that all non-linear dynamic systems with more than two degrees of freedom can display unpredictable (chaotic) behaviour over a prolonged period. Such systems can have multiple stable states governed by 'equilibrium', 'limit cycle' or 'chaotic'('strange') attractors.

The human brain is a rather superior information processor compared to any artificial system devised so far. Brain modelling comprises a detailed study of the different processes involved in biological neural computation and the incorporation of the facts so learnt into more realistic models of brain function, thus making them more 'intelligent', in comparison with presently available systems in performing cognitive tasks. There have been artificial neural networks of excitatory-inhibitory neural pairs exhibiting chaotic behaviour in certain parameter regions. Hence a complete knowledge of the domains of sub-harmonic behaviour and chaos in the parameter space is very important. Chaotic systems are more flexible than non-chaotic ones since the attractor spans a large volume of the state space, and, with proper control, one can rapidly switch among many different behaviours. Using variable feedback control, the networks can be made to converge to any possible periodic pattern. On withdrawal of the control signal, the network reverts to the chaotic state. Similarly, the phenomenon of state synchronisation among elements of a coupled chaotic network has also been studied, emulating the neural basis of 'attention'.

Since electroencephalographic (EEG) signals may be considered chaotic, non-linear dynamics and deterministic chaos theory may supply effective quantitative descriptors of EEG dynamics and of the underlying chaos in the brain. Karhunen-Loeve decomposition of the covariance matrix of the EEG signal has been used to analyse EEG signals of 4 healthy subjects, under drug-free condition and under the influence of diazepam. KL-complexity of the signal differs profoundly for the signals registered in different EEG channels, from about 5-8 for signals in frontal channels up to 40 and more in occipital ones. But no consistency in the influence of diazepam administration on KL-complexity is observed. The embedding dimension of the EEG signals of the same subjects was estimated, which turned out to be between 7 and 11, thereby supporting the presumption that there is a low-dimensional chaotic attractor. Therefore, non-linear time series analysis can be used to investigate the dynamics underlying the generation of EEG signals. This approach does not seem practical yet, but deserves further study.

As of now, EEG signals are considered to be generated by non-linear dynamic systems exhibiting chaotic behaviour. The system may behave as a deterministic chaotic attractor. The complexity of the attractor can be characterised by the correlation dimension that can be computed from one signal generated by the system. The non-linear properties of biological neural networks are at the threshold range for spike generation and saturation phenomena. Also, these populations interact by means of feedback loops with time delays. For calculating the correlation dimension, the EEG signal to be analysed comprises a set of samples x_n, defined from $n = 1, ..., N+k$. From this series, one can construct m-dimensional vectors (in an N-dimensional space) $V_m(i)$ with $i = 1,...,N$, defined by :

$$V_m(i) = (x(i+k_0),\ x(i+k_1), ..., x(i+k_{m-1})) \tag{6.1}$$

using a fixed set of delays k_l with $0 \le k_l \le k$ $(l=0,...,m-1)$ and $k_l \neq k_p$ if $p \neq 1$. The correlation integral $C(r,m)$, where r is the radius of a sphere with centre $V_m(i)$ in R^m may be defined as

$$C(r,m) = 1/N^2 \sum_{i=1}^{N-1} \sum_{j=i+1}^{N} h(r-d)(V_m(i), V_m(j)) \tag{6.2}$$

where h is the Heaviside or step function ($h(x < 0) = 0$), $h(x \geq 0) = 1$), d is the distance between 2 vectors, for which one may use the largest difference between the corresponding components.

If an attractor is present in the time series, then, the values $C(r,m)$ should satisfy, for small r and large m and N: $C(r,m) \propto r^{D_2}$, where D_2 is the correlation dimension of the attractor and is given by the slope of the plot of log $C(r,m)$ versus log(r). $C(r,m)$ is calculated as per Grassberger and Procaccia. During epileptic seizures D_2 is found to be low (between 2 and 4).

The Lyapunov spectrum (Ikeguchi et al. 1990) of EEG has also been analysed. The computation of the full Lyapunov spectrum requires considerably more effort than just the maximal exponent. An essential ingredient is some estimate of the local Jacobians, i.e., of the linearised dynamics, which rules the growth of infinitesimal perturbations. One either finds it from direct fits of local linear models of the type $a_{n+1} = a_n s_n + b_n$ such that the first row of the Jacobian is the vector a_n, and $(J)_y = d_{i-1}$, for $i=1, ..., m$, where m is the embedding dimension. The a_n is given by the least squares minimisation $\alpha^2 = \Sigma_1 (s_{i+1} - a_n s_i - b_n)^2$, where $\{a_i\}$ is the set of neighbours of s_n. Or one constructs a global non-linear model and computes its local Jacobians by taking derivatives. In both cases, one multiplies the Jacobians one by one, following the trajectory, to as many different vectors u_k in tangent space as one wants to compute Lyapunov exponents. Every few steps, one applies a Gram-Schmidt orthonormalisation procedure to the set of u_k, and accumulates the logarithms of their re-scaling factors. Their average, in the order of the Gram-Schmidt procedure, gives the Lyapunov exponents in descending order. The routine using this method employs local linear fits. Apart from the problem of spurious exponents, this method contains some other pitfalls: It *assumes* that there exist well-defined Jacobians, and does not test for their relevance. In particular, when attractors are thin in the embedding space, some (or all) of the local Jacobians might be estimated very badly. Then the whole product can suffer from these bad estimates and the exponents are correspondingly wrong. Thus the global non-linear approach can be superior, if a modelling has been successful.

The computation of the first part of the Lyapunov spectrum allows for some interesting cross-checks. It has been conjectured, and is found to be correct in most physical situations, that the Lyapunov spectrum and the fractal dimension of an attractor are closely related. If the expanding and least contracting directions in space are continuously filled and only one partial dimension is fractal, then one can ask for the dimensionality of a (fractal) volume such that it is invariant, i.e. such that the sum of the corresponding Lyapunov exponents vanishes, where the last one is weighted with the non-integer part of the dimension:

$$D_{ky} = k + \frac{\sum_{i=1}^{k} \lambda_i}{|\lambda_{k+1}|}$$

where k is the maximum integer such that the sum of the k largest exponents is still non-negative. D_{ky} is conjectured to coincide with the information dimension.

The Pesin identity is valid under the same assumptions and allows computing the KS-entropy:

$$h_{ks} = \sum_{i=1}^{m} \theta(\lambda_i)\lambda_i$$

6.6 Further Reading

[1] Sarbadhikari, S. N. and Pal, S.K. Automated techniques for identifying depression from EEG. pp. 51–82, chapter 3, vol. 4. In *Handbook of Computational Methods in Biomaterials, Biotechnology & Biomedical Systems.* Ed. C.T. Leondes. Kluwer Academic Publishers. 2002.

[2] Sarbadhikari, S. N. and Chakrabarty, K. Chaos in the brain: A short review alluding to epilepsy, depression, exercise and lateralization. *Med. Eng. Phys.* 23(7): 445–455. 2001.

[3] Glass, L. Synchronization and rhythmic processes in physiology. *Nature* 410: 277–284. 2001.

[4] Thomasson, N., Pezard, L., Allilaire, J-F., Renault, B. and Martinerie J. Nonlinear EEG changes associated with clinical improvement in depressed patients. *Nonlinear Dynamics, Psychology & Life Sc.* 4: 203–218. 2000.

[5] Melancon, G., Joanette, Y., Belair, J. Chaos, brain, and cognition: Toward a nonlinear order? *Brain Cogn.* 42:33–36. 2000

[6] Moss, F. and Gielen, S. Eds. *Handbook of Biological Physics* vol.4. *Neuroinformatics.* Amsterdam: Elsevier. 2000.

[7] Lopes da Silva, F. EEG analysis: Theory and practice. pp. 1135–1163, ch. 61. In *Electroencephalography – Basic Principles, Clinical Applications & Related Fields.* 4th ed. Ed. E. Niedermeyer & F. Lopes da Silva. Baltimore: Williams & Wilkins. 1999.

[8] Schiff, S. J. Forecasting brain storms. *Nature Med.* 4:1117–1118. 1998

[9] Mitra, S. K. and Sarbadhikari, S. N. Iterative function system and genetic algorithm based EEG compression. *Med. Eng. Phys.* 19(7): 605–617. 1997.

[10] Aine, C., Okada, Y. C., Stroink, G., Swithenby, S. J. and Wood, C. C. Eds, *Advances in Biomagnetism : Biomag 96.* Springer Verlag. 1997.

[11] Stam, C.J., Jelles, B., Achtereekte, H.A., van Birgelen, J.H., Slaets, J.P. Diagnostic usefulness of linear and nonlinear quantitative EEG analysis in Alzheimer's disease. *Clin Electroencephalogr* 27:69–77. 1996

[12] Mitra, S. and Pal, S. K., Fuzzy multi-layer perceptron, inferencing and rule generation. *IEEE Trans. Neural Networks* 6: 51–63. 1995

[13] Nunez, P. L. Ed. *Neocortical Dynamics and Human EEG Rhythms.* New York: Oxford University Press. 1995.

[14] Duffy, F. H. The role of quantified EEG in psychological research. pp. 93–132. In *Human Behavior and the Developing Brain.* Eds. G. Dawson and K.W. Fischer. Guilford Press. 1994.

[15] Bezdek, J. C. and Pal, S. K. *Fuzzy Models for Pattern Recognition.* IEEE Press. New York. 1992.

[16] Yan, Y. Nunez, P. L. and Hart, R. T. Finite-element model of human head: Scalp potentials due to dipole sources. *Med. Biol. Eng. Comput.* 29: 475–481. 1991.

[17] Cesarelli, M., Clemente, F. and Bracale, M., The flexible FFT algorithm for processing biomedical signals using a PC. *J. Biomed. Eng.* 12: 527–530. 1990

[18] Ikeguchi, T. Aihara, K., Itoh, S. and Utsunomiya, T. An analysis on Lyapunov spectrum of EEG potentials. *Trans. Inst. Electron. Inf. Commun. Eng. E.* E73(6): 842– 47. 1990.

[19] Ning, T. and Bronzino, J. D. Autoregressive and bispectral analysis techniques: EEG applications. *IEEE Eng. Med. Biol.* 9: 47–49. 1990.

[20] Skarda, C. A. and Freeman, W. J. How brains make chaos in order to make sense of the world. *Behav. Brain Sci.* 10: 161–195. 1987.

[21] Gevins, A. S. Analysis of the electromagnetic signals of the human brain: Milestone, obstacles and goals. *IEEE Trans. Biomed. Eng.* 31: 833–835. 1984

[22] Grassberger, P. and Procaccia, I. Measuring the strangeness of strange attractors. *Physica.* 9D: 189–208. 1983.

[23] Hjorth, B. Time domain descriptors and their relation to a particular model for generation of EEG activity. pp. 3–8. In *CEAN—Computerized EEG Analysis*. Eds. G.Dolce and Künkel. Stuttgart: Fischer. 1975.

[24] Kellaway, P. and I. Petersen. Eds. *Automation of Clinical Electroencephalography*. New York: Raven Press. 1973.

[25] Cooley, J. W. and Tukey, J. W. An algorithm for the machine calculation of Complex Fourier Series. *Math. Comput.* 19: 297–301. 1965

6.7 Questions

1. Descriptive Questions

1. Comment on the statement: Digital signal processing often replaces or supplements signal analysis in the continuous time domain.
2. What are the non-parametric methods for EEG analysis?
3. What are the parametric methods for EEG analysis?
4. How is wavelet transform comparable to WDFT ?
5. What are the non-linear methods for EEG analysis?

2. Write Short Notes on

1. FFT
2. Power (density) spectrum
3. ARMA
4. Wavelet analysis
5. Bispectrum
6. Cepstrum
7. Bicepstrum
8. BEAM
9. Correlation dimension
10. Running fractal dimension

3. Multiple Choice Questions

Choose the single best response.

1. A typical bioelectrical signal like EEG may be analysed
   (A) parametrically
   (B) non-parametrically
   (C) spatially
   (D) non-linearly
   (E) any of the above

2. FFT is also known as power (density) spectrum because
   (A) it is a powerful tool
   (B) amplitude is squared
   (C) density of the power in EEG is measured
   (D) none of the above

3. Convolution means
   (A) revolution
   (B) convexity
   (C) superimposition while stepping sequentially
   (D) none of the above

4. Whether a chaotic system will find itself in any stable state depends on the
   (A) system's parameters
   (B) input conditions
   (C) both of the above
   (D) none of the above

5. Electrophysiological signal analysis consists of
   (A) feature extraction
   (B) classification
   (C) pattern recognition
   (D) all of the above

Answers

1 E; 2 B; 3 C; 4 C; 5 D

7

Methods of Medical Data Acquisition, Analysis, Compression and Monitoring

7.1. Introduction
7.2. Analog instruments
7.3. Digitisation
7.4. Pre-amplification
7.5. Filtering
7.6. Compression
7.7. Monitoring
7.8. Further reading
7.9. Questions

7.1 Introduction

A *data acquisition control system* relates to a computer system in which high speed data channels are used to connect a central computer to distant locations (e.g., for the purpose of telemedicine). It is a system devised to operate in real time applications, receiving data from many remote stations and transmitting data to them under program control. The data acquisition devices in such a system may cater to both analog and digital forms of data input and output. For online analysis of data, cathode ray tube (CRT) video display units (VDUs) form the data analysis display unit.

Digital signal processing (DSP) is concerned with the theoretical and practical aspects of representing information bearing signals in digital form, and with using computers or special purpose digital hardware either to extract that information or to transform the signals into useful forms. Areas where DSP has made a significant impact include telecommunications, man–machine communications, and medical technology, use of radar and sonar, and seismic data analysis. It is the basis of many areas of technology, from mobile phones to modems and multimedia PCs. DSP devices account for a significant proportion of worldwide semiconductor sales, amounting to billions of dollars annually and major electronics manufacturers have invested heavily in DSP technology.

DSP is traditionally a highly mathematical subject, and standard DSP textbooks contain a lot of mathematical exposition. This is necessary for a profound understanding of the subject. However, many people prefer a simple introduction to the basic concepts and techniques of DSP. Please visit the website, www.techonline.com, to learn more about DSP.

Signals commonly need to be processed in a variety of ways. For example, the output signal from a transducer may well be contaminated with unwanted electrical 'noise'. The electrodes attached to a patient's chest when an ECG is taken, measure tiny electrical voltage changes due to the activity of the heart and other muscles. The signal is often strongly affected by 'mains pickup' due to electrical interference from the mains supply. Processing the signal using a filter circuit can remove or at least reduce the unwanted part of the signal. Increasingly, nowadays, the filtering of signals to improve signal quality or to extract important information is done by DSP techniques rather than by analog electronics.

The development of digital signal processing dates from the 1960s with the use of mainframe digital computers for number-crunching applications such as the fast Fourier transform (FFT, see Section 6.3.1), which allows the frequency spectrum of a signal to be computed rapidly. These techniques were not widely used at that time, because suitable computing equipment was available only in universities and other scientific research institutions.

The introduction of the microprocessor in the late 1970s and early 1980s made it possible for DSP techniques to be used in a much wider range of applications. However, general-purpose microprocessors such as the Intel x86 family are not ideally suited to the numerically-intensive requirements of DSP, and during the 1980s the increasing importance of DSP led several major electronics manufacturers to develop digital signal processor chips—specialised microprocessors with architectures designed specifically for the types of operations required in digital signal processing. (Note that the acronym DSP can variously mean digital signal processing, the term used for a wide range of techniques for processing signals digitally, or a digital signal processor, a specialised type of microprocessor chip). Like a general-purpose microprocessor, a DSP is a programmable device, with its own native instruction code. DSP chips are capable of carrying out millions of floating point operations per second, and like their better-known general-purpose cousins, faster and more powerful versions are continually being introduced. Because they now find application in mass-market products, DSP chips account for a substantial proportion of the world market for electronic devices. Sales amount to billions of dollars annually, and seem likely to continue to increase rapidly. Now ASIC chips or 'application specific integrated circuits' are also available for specific uses.

An important application of DSP is in signal compression and decompression. In CD systems, for example, the music recorded on the CD is in a compressed form (to increase storage capacity) and must be decompressed for the recorded signal to be reproduced. Signal compression is used in digital cellular phones to allow a greater number of calls to be handled simultaneously within each local 'cell'. DSP signal compression technology allows people not only to talk to one another by telephone but also to see one another on the screens of their PCs, using small video cameras mounted on the computer monitors, with only a conventional telephone line linking them together.

Although the mathematical theory underlying DSP techniques such as fast Fourier and Hilbert transforms, digital filter design and signal compression can be fairly complex, the numerical operations required to implement these techniques are, in fact, very simple, consisting mainly of operations that could be done on a cheap four-function calculator. The architecture of a DSP chip is designed to carry out such operations incredibly fast, processing up to tens of millions of samples per second, so as to provide real time performance: that is, the ability to process a signal 'live' as it is sampled and then output the processed signal, for example to a loudspeaker or video display. All the practical examples of DSP applications mentioned earlier, such as hard disc drives and mobile phones, demand real time operation.

7.2 Analog Instruments

Analog instruments may be used for acquiring biomedical signals and images. They may then be digitised, or direct digital recorders may be used. Direct analysis of analog signals like ECG is possible, whereas in most cases of EEG, it is neither easy nor reliable. Therefore, digitisation and further processing for analysis is often essential. The process of digitisation and display has been described in Chapter 5.

While compressing images/signals for transmission, reproducibility or fidelity must be carefully considered. There should be a high signal-to-noise ratio (SNR) – otherwise after transmission and reconstruction, the corresponding faulty image/signal might give rise to an entirely incorrect or erroneous diagnosis. Therefore, compression algorithms must be simple, easily to implement, and at the same time, reliable (high SNR). If the SNR is high, it implies that the reconstructed signal or image will be almost as good as the original. This is termed as *fidelity* of the signal or image (so now you realize that your hi-fi stereo system is one which reproduces the soundtrack as well as the original recording quality, which means it has *high fi*delity!).

Two error metrics used to compare the various image compression techniques are the mean square error (MSE) and the peak signal-to-noise ratio (PSNR). The MSE is the cumulative squared error between the compressed and the original image, whereas PSNR is a measure of the peak error. The mathematical formula for MSE is:

$$\frac{1}{MN}\sum_{y=1}^{M}\sum_{x=1}^{N}[I(x,y)-I'(x,y)]^2$$

While, for PSNR it is: $20\times\log 10\left(\frac{255}{\sqrt{\text{MSE}}}\right)$

where $I(x,y)$ is the original image, $I'(x,y)$ is the approximated version (which is actually the decompressed image) and M,N are the dimensions of the images. A lower value for MSE means lesser error, and as seen from the inverse relation between the MSE and the PSNR, this translates to a high value of PSNR. Logically, a higher value of PSNR is good because it means that the ratio of signal-to-noise is higher. Here, the 'signal' is the original image, and the 'noise' is the error in reconstruction. So, a compression scheme having a lower MSE (and a high PSNR), is a better one.

For monitoring biomedical signals, analog, as well as, digital systems are in vogue.

Medical amplifiers, designed for OEM applications, are complete, low power analog signal processing amplifier modules which contain all the necessary signal conditioning elements for processing many different biological signals. These units are high performing biomedical amplifiers, and are configurable for any signal from the body (human or animal).

They are usually:

i. immune to strong magnetic fields such as those in magnetic resonance imaging (MRI)
ii. immune to radio frequency interference (RFI)
iii. immune to mechanical motion of electrode leads

Applications for most types of medical amplifiers include:

i. analog signal conditioning for bioelectrical signals

ii. DSP analog signal processing
iii. EEG signal processing
iv. EMG signal processing
v. EOG signal processing
vi. EKG signal processing
vii. Respiration sensor signal processing
viii. Oxygen sensor signal processing
ix. Strain gauge signal processing
x. Biomedical signal ambulatory recorders

7.3 Digitisation

To analyse digitised data, it must be quantified. This will help to give an objective impression and the data will not be user-dependent. There will also be more intra-observer and inter-observer concurrence (i.e., less variability). Chapter 6 deals with some quantification techniques.

In many cases, the signal is initially in the form of an analog electrical voltage or current, produced for example by a microphone or some other type of transducer. In some situations, the data is already in digital form – such as the output from the readout system of a CD (compact disc) player. An analog signal must be converted into digital (i.e., numerical) form before DSP techniques can be applied. An analog electrical voltage signal, for example, can be digitised using an integrated electronic circuit (IC) device called an analog-to-digital converter or ADC. This generates a digital output in the form of a binary number whose value represents the electrical voltage input to the device.

7.4 Pre-amplification

For the purpose of recording and analysing biomedical signals, pre-amplification of the signals is absolutely necessary, since all biopotentials are too small (magnitude in microvolts) for detection. As bandwidth demand rises, the construction of optical packet-switching nodes targeting optical routers would benefit from fast optical switches. Semiconductor-optical-amplifier (SOA) technology provides this high-speed switching capability as well as high gain, high extinction ratio, and high integration potential. Moreover, it is a key technology for several other functions, including all-optical wavelength conversion, regeneration, wavelength selection, booster and in-line amplification, in-node optical pre-amplification, and mid-span spectral inversion. Also please see Section 1.5.

7.5 Filtering

In signal processing, the function of a filter is to remove unwanted parts of the signal, such as random noise, or to extract useful parts of the signal, such as the components lying within a certain frequency range. There are two main kinds of filter, analog and digital. They are quite different in their physical makeup and in how they work.

An *analog filter* uses analog electronic circuits made up from components such as resistors, capacitors and op amps to produce the required filtering effect. Such filter circuits are widely used in such applications as noise reduction, video signal enhancement, graphic equalizers in hi-fi systems, and many other areas.

There are well-established standard techniques for designing an analog filter circuit for a given requirement. At all stages, the signal being filtered is an electrical voltage or current which is the direct analogue of the physical quantity (e.g., a transducer output) involved.

A *digital filter* uses a digital processor to perform numerical calculations on sampled values of the signal. The processor may be a general-purpose computer such as a PC, or a specialised DSP chip. The analog input signal must first be sampled and digitised using an ADC (analog-to-digital converter). The resulting binary numbers, representing successive sampled values of the input signal, are transferred to the processor, which carries out numerical calculations on them. These calculations typically involve multiplying the input values by constants and adding the products together. If necessary, the results of these calculations, which now represent sampled values of the filtered signal, are output through a digital-to-analog converter (DAC) to convert the signal back to analog form. Please note that in a digital filter, the signal is represented by a sequence of numbers, rather than a voltage or current. Also please see Section 5.3.

7.6 Compression

The need for signal and image compression is ubiquitous in medical images and signals (still and video) processing. Where data stores become very large (e.g., video, store and forward medical signals and images), the need is not met by simple (loss-less) file compression schemes, and sophisticated (lossy) coding techniques become necessary. The common modes are loss-less entropy based coding including Huffman and Lempel-Ziv coding, and lossy compression techniques including scalar quantisers, transform coding (Karhunen–Loeve, discrete cosine transform (DCT), and non-linear transform codes), predictive coding, vector quantisation, adaptive codes, iterative function systems or fractals and genetic algorithm (IFS-GA) for fast search of self-similarities in the given signal or image [2] and wavelets (Section 6.3.2). The relation between compression schemes and probabilistic data modelling has to be emphasised in conjunction with each technique. The prerequisites for actually doing these are: undergraduate calculus, introductory probability and statistics, and, some programming experience.

The compressed image or signal is transmitted to the target end. There it has to be decompressed by performing inverse operations to get back the original signal.

The common formats of file compression commercially available now are JPG (JFIF), GIF, and TIFF which use various compression schemes like JPEG, LZW, G3/G4 and CMP. For medical images especially, digital imaging and communications in medicine (DICOM) is the standard, which supports retrieval of information from imaging devices/equipment to diagnostic and review workstations, and to short-term and long-term storage systems.

For proper compression, research on the theory, algorithm design, simulation, and quality evaluation of systems for signal compression, quantisation, source coding with a fidelity criterion, and classification are essential. Related topics in signal processing also have to be considered, including enhancement, segmentation, regression, detection, estimation, prediction, video coding, and motion estimation. A common theme of the work is the interplay of compression and classification, especially of tree-structured algorithms and clustering for vector quantisation for simultaneous compression and classification.

A novel technology, based on the modified Set partitioning in hierarchical trees (SPIHT) technology, makes it possible to compress natural images (including video) and signals (including voice) with the given level of losses (which will not damage the information) to an extremely good compression ratio.

This feature is very important for telecommunication systems with limited time and/or band. The main difference between this technology and the original SPIHT technology is in the use of modified wavelet transform and corresponding change in coding. This makes it possible to obtain better compression rates on an average and in lesser time.

Another novel technology, based on S-transform, a particular modification of wavelet transform, makes it possible to compress natural images (including video) and signals (including voice) with no losses and a very good compression ratio.

7.7 Monitoring

Patient monitors measure, monitor, analyse and access patient data with incredible accuracy. After imaging, non-invasive physiological measurement in one form or another is one of the most important diagnostic methodologies used in health care. It includes ECG, EMG and other bioelectric measurements like non-invasive blood pressure reading, pulse oximetry, pulse rate and temperature reading. It underpins patient monitoring and treatment in the intensive care unit (ICU), the operating theatre during anaesthesia, and is an integral part of services such as audiology.

Future developments in this field will have an enormous impact on the development of information technology in medicine, especially telemedicine. Thus, the benefits of research into these techniques are very widespread in health care and in applied physiology in general.

Any site in a hospital requires maximum flexibility and optimum monitoring. Intelligent monitoring, intelligent alarming, plug and play modules, transmission control protocol/internet protocol (TCP/IP), ethernet networking and many other features provide easy, integrated monitoring in any facility. Some of the components of a monitoring system are:

i. central monitors
ii. telemetry systems
iii. bedside stations – the monitoring system can be used as a central or bedside monitor
iv. modular, full-featured, compact bedside monitor
v. multi-parameter bedside monitors with ECG management
vi. portable bedside monitor with easy touch screen operation
vii. multi-gas unit
viii. telemetry transmitters for monitors
ix. remote monitoring software for Windows PC

Occasionally there could also be:

i. pocket-sized SpO_2/CO_2 monitors
ii. pulse oximeters (see Section 16.3.1)
iii. dye densitogram analysers – they measure blood volume, cardiac output and other circulatory parameters

Despite their usefulness, monitors often seem to be quite frightening because of the sheer size of their control panels. Nevertheless, now most manufacturers make them user-friendly with mostly menu-driven systems, easily visible and operable switches, explanatory printouts, proper alarm signalling, and for all practical purposes fail-safe operability.

7.8 Further Reading

[1] Raja Rao, C and Guha, S. K. *Principles of Medical Electronics and Biomedical Instrumentation.* Hyderabad, India: Universities Press Pvt. Ltd. 2001.

[2] Mitra, S. K. and Sarbadhikari, S. N. Iterative function system and genetic algorithm based EEG compression. *Med. Eng. Phys.* 19(7): 605–617. 1997.

[3] Gonzalez, R. C. and Woods, R. E. *Digital Image Processing.* Reading, MA: Addison-Wesley. 1992.

[4] Oppenheim, A. V. and Schafer, R. W. *Digital Signal Processing.* New Delhi: Prentice-Hall of India Pvt. Ltd. 1989.

[5] Stearns, S. D. and David, R. A. *Signal Processing Algorithms.* Englewood Cliffs, New Jersey: Prentice-Hall Inc. 1988.

[6] Jay, F. *IEEE Standard Dictionary of Electrical and Electronics Terms.* New York: IEEE Inc., John Wiley & Sons. 1977.

7.9 Questions

1. Descriptive Questions

1. What are the prerequisites for the analysis of digitised data?
2. What should be the ideal characteristics for image and signal compression?

2. Write Short Notes on

1. Digital signal processing
2. User-friendliness
3. Fidelity
4. SNR
5. CRT

3. Multiple Choice Questions

Choose the single best response.

1. DSP refers to
   (A) digital signal processing
   (B) digital signal processor
   (C) either
   (D) none
2. For high fidelity, SNR should be
   (A) high
   (B) low
   (C) one
   (D) none
3. For telemedicine, a medical image from the originating centre to the target centre should undergo
   (A) acquisition-digitisation-filtering-decompression

(B) acquisition-digitisation-compression-transmission-decompression
(C) acquisition-filtering-compression-digitisation
(D) acquisition-decompression-compression-transmission

4. A patient monitoring system may be used in the
   (A) intensive care unit
   (B) operation theatre
   (C) critical care unit
   (D) any of the above

5. A notch filter (in India) is commonly of cut-off frequency
   (A) 50 Hz
   (B) 60 Hz
   (C) 70 Hz
   (D) 80 Hz

Answers

1. C 2. A 3. B 4. D 5. A

Section 3

Biomechanics

Biomechanics is the science of forces acting internally or externally on the living organism or the cell. It may also be defined as the application of mechanical laws to living structures, as to a locomotor system. The relationship between the behaviour of biological structures and physical influences of a biomaterial or prosthesis is also studied under this science.

Chapters 13, and 15 deal with biomaterials. Chapter 8 deals with the mechanical properties of bones like viscoelasticity, anisotropy, yield strength, and also models of bones including FEM and composite models.

Chapter 9 deals with the biomechanics of implant tissues and tissue engineering while Chapter 10 discusses sports biomechanics

8

Properties and Models of Bones

8.1 Introduction

The *bone* is a type of connective tissue where the matrix is hard (compared to other soft tissues of the body), and calcified (contain calcium in the form of a mineral – apatite) and has a rather complex structure. The matrix contains many collagen fibre layers. The term for these layers of the matrix is *lamellae*.

Bones may be classified into: cortical or compact bone (e.g., long bones) and cancellous, spongy, porous or trabecular bone (e.g., sternum). Compact bones usually comprise the perforating (Volkmann's) canal, the periosteum (dense connective tissue covering over the bone) and the Haversian system (osteon). The diaphysis is the shaft of the long bone, while the epiphysis is at the ends of the bone (Fig. 8.1). The articular cartilage is a form of hyaline cartilage, which covers the epiphyseal surface where a bone forms a joint with another bone. Epiphyseal surfaces which are not part of an articulating surface are covered by the periosteum. The markings on bones (such as tuberosities, lines, crests, and spines) serve as sites of attachment for muscles and/or ligaments. The openings in bones (such as foramen, fossa, fissures, etc.) serve as passageways for blood vessels and/or nerves. The projections on bones (like heads and condyles) take part in forming joints.

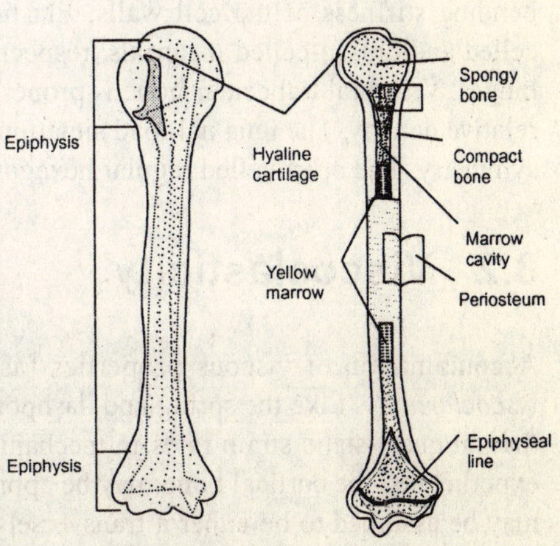

Fig. 8.1 Structure of a long bone

The spongy bone has no Haversian systems; instead it has a web-like arrangement (trabeculae).

The cells in the bones are: osteocytes, osteoblasts (cells that lay down new bone) and osteoclasts (cells that destroy old bone so that growth can occur). Osteoclasts also begin to destroy bone when the body is deficient of calcium and when the bone is not used.

Another classification of bones may be:

(a) ***Long bones*** These are made primarily of compact bone, except at the ends (epiphysis) which has only a thin layer of compact bone covering a great deal of spongy bone. Examples of long bones include the humerus and femur. They are bones that are longer than they are wide;
(b) ***Short bones*** These are bones shaped like the carpals and tarsals, and contain mostly spongy bone;
(c) ***Flat bones*** Examples of flat bones are those of the skull. They contain spongy bone in between surrounding layers of compact bone;
(d) ***Irregular bones*** Examples are the vertebrae, and,
(e) ***Sesamoid bones*** These, like the patella, are enclosed in tendons.

Cortical bone is a very dense, tough material that provides great stiffness and strength in a relatively compact structure. Trabecular bone, on the other hand, is a porous, sponge-like network of bone material. The structure of a trabecular bone adapts to the loading environment in such a way as to maximize the structural stiffness and strength while minimising the weight. It follows that the symmetry of the structure will depend on the loading situation. It is well known, for example, that the femur and tibia have trabecular bone that is highly asymmetric in order to support the existing multi-axial state of stress. The vertebrae of the spine, in contrast, are loaded primarily in one direction (axial), so that the trabecular bone of the vertebra often develops into a symmetric arrangement of vertically oriented columns with reinforcing horizontal struts.

Trabecular bone is generally characterised as a cellular solid, or foam, consisting of an interconnected network of rods and plates. The mechanical behaviour of foams can be described by analysing the mechanisms by which the cells deform. The overall behaviour depends on three governing parameters: the type of cell structure (open- or closed-celled); the volume fraction of solids, or relative density; and the properties of the cell wall material. Generally, the elastic moduli will depend on the axial and bending stiffness of the cell walls. Elastic collapse is caused by beam or plate buckling for open-celled and closed-celled materials, respectively. Plastic collapse is caused by the formation of plastic hinges. Vertebral trabecular bone is properly treated as an open-celled material, since it has a very low relative density. The uniaxial load situation dictates that the structure will be columnar with cylindrical symmetry. The open-celled regular hexagon is therefore chosen as the appropriate unit cell.

8.2 Viscoelasticity

A combination of viscous properties (as in fluids) and elastic properties (as in solids) is called *viscoelasticity*. Like the spring and dashpot models, bones also exhibit viscoelastic properties. Despite that, at quasi-static strain rates in mechanical testing, and also in ultrasonographic frequency used in experiments, the cortical bone may be approximately modelled as an anisotropic linear elastic solid. It may be assumed to be either a transversely isotropic or an orthotropic material. Plexiform or laminar bones have higher compressive and shear anisotropy factors and are, therefore, stiffer, more rigid and anisotropic, compared to Haversian bones. The symmetrical arrangement of a structure in which there is a unique axis perpendicular to a plane in which the other two axes are equivalent, is known as *transverse isotropy*. The direction of the long bone is chosen as the unique axis.

For viscoelastic materials like bone, Hooke's law for linear elastic material has to be replaced with a constitutive equation that includes time dependence of material properties. This sort of an anisotropic linear viscoelastic material may be described by the Boltzmann superposition intergral.

$$\sigma_{ij}(t) = \int_{-\infty}^{t} C_{ijkl}(t-\tau)\frac{d\varepsilon_{kl}(\tau)}{d\tau}d\tau \tag{8.1}$$

where $s_{ij}(t)$ and $e_{kl}(t)$ are the time-dependent second rank stress and strain tensors, respectively, and C_{ijkl} $(t–t)$ is the fourth rank relaxation modulus tensor. This tensor has 36 independent elements for the lowest symmetry case and 12 non-zero independent elements for an orthotropic (symmetrical arrangement of a structure where there are three distinct orthogonal axes of symmetry) solid. If we apply equation (8.1) to an orthotropic material (e.g., plexiform bone), for uniaxial tension in direction 1, we get:

$$\sigma_{11}(t) = \int_{-\infty}^{t}\left[C_{11}(t-\tau)\frac{d\varepsilon_{11}(\tau)}{d\tau} + C_{12}(t-\tau)\frac{d\varepsilon_{22}(\tau)}{d\tau} + C_{13}(t-\tau)\frac{d\varepsilon_{33}(\tau)}{d\tau}\right]d\tau \tag{8.2}$$

$$\sigma_{22}(t) = \int_{-\infty}^{t}\left[C_{21}(t-\tau)\frac{d\varepsilon_{11}(\tau)}{d\tau} + C_{22}(t-\tau)\frac{d\varepsilon_{22}(\tau)}{d\tau} + C_{23}(t-\tau)\frac{d\varepsilon_{33}(\tau)}{d\tau}\right]d\tau = 0 \quad \text{(for all } t\text{)} \tag{8.3}$$

$$\sigma_{33}(t) = \int_{-\infty}^{t}\left[C_{31}(t-\tau)\frac{d\varepsilon_{11}(\tau)}{d\tau} + C_{32}(t-\tau)\frac{d\varepsilon_{22}(\tau)}{d\tau} + C_{33}(t-\tau)\frac{d\varepsilon_{33}(\tau)}{d\tau}\right]d\tau = 0 \quad \text{(for all } t\text{)} \tag{8.4}$$

Making the integrands vanish purvey an obvious solution to equations (8.3) and (8.4). Solving them simultaneously for $\frac{d\varepsilon_{22}(\tau)}{d\tau}$ and $\frac{d\varepsilon_{33}(\tau)}{d\tau}$, and substituting these values in equation (8.1) yields:

$$\sigma_{11}(t) = \int_{-\infty}^{t} E_{11}(t-\tau)\frac{d\varepsilon_{11}(\tau)}{d\tau}d\tau$$

where, for convenience, denoting $C_{ij} \equiv C_{ij}\,(t-t)$, Young's modulus is:

$$E_{11}(t-\tau) = C_{11} + C_{12}\frac{[C_{31} - (C_{21}C_{33}/C_{23})]}{[(C_{21}C_{33}/C_{23}) - C_{32}]} + C_{13}\frac{[C_{21} - (C_{31}C_{22}/C_{32})]}{[(C_{22}C_{33}/C_{32}) - C_{23}]} \tag{8.5}$$

Here, for uniaxial tension (compression), only 9 independent orthotropic tensor components are involved, the 3 shear components being zero. Now, if we assume that the bone is being driven by a strain at a frequency w, with a corresponding sinusoidal stress lagging by an angle d, then the complex Young's modulus $E^*(w)$ may be denoted by:

$$E^*(\omega) = E'(\omega) + iE''(\omega) \tag{8.6}$$

where $E'(\omega)$ (stress–strain ratio in phase with the strain) is the storage modulus and $E''(\omega)$ (stress–strain ratio, 90° out of phase with the strain) is the loss modulus. The ratio of the loss modulus to the storage modulus is equal to tan*d*. Commonly, data are depicted by the graph of the storage modulus along with a graph of tan*d*, both against frequency.

Yield strength/point is defined as the stress beyond the elastic limit that results in permanent bending, deflection, or deformation (i.e., the metal goes from exhibiting pure elastic behaviour to a combination of elastic/plastic elongation). At the offset yield point the sample will have deformed plastically to an extent that will leave the material with a permanent strain of 0.002 (0.2%) when it is at zero stress. The stress, at which this occurs, is the *off set yield stress*. For metals, the elastic limit is defined as the 0.2% offset yield strength. This represents the stress at which the stress-train curve for uniaxial tensile loading deviates by a strain of 0.2% from the linear-elastic line. It is the same in tension and compression. For polymers, the elastic limit is the stress at which the uniaxial stress-strain curve becomes markedly non-linear: typically, a strain of 1%. This may be caused by *shear yielding* (irreversible slipping of molecular chains) or by *crazing* (formation of low density, crack-like volumes which scatter light, making the polymer look white). For fine ceramics and glasses, the elastic limit is an estimate, based on the tensile strength (which is low due to brittle fracture). When based on direct measurements at high pressures, or on hardness measurements, the stress required to cause plastic flow, is very high; higher than the compressive strength, which is lowered by crushing. For composites, the elastic limit is best defined by a set deviation from linear-elastic uniaxial behaviour: e.g., 0.5%. The *average yield strength* is the strength which induces a specific permanent deformation.

Mechanical stiffness values have been measured most often for secondary haversian bone human and cow (bovine) bone as well as bovine plexiform bone. Stiffness has been measured using both standard mechanical testing techniques as well as ultrasonic measurements where the velocity of waves propagating through a material is measured. This wave velocity is related to the stiffness and the density of the material. Such testing have produced no cage deformation i.e, deformation of the outer structure (cage) of the bone, at the stress levels applied, except some elastic deformation. Bone itself, i.e., young adult bone, begins to break at about 9,600 Newtons (N). (A mnemonic to remember this: there are four apples to the pound; there are 4N to the pound. Divide 9600 N by 4.4 to tell how many pounds were applied.) Please also see Section 13.3. Studies indicate that changes in failure energy (the energy required to stress a bond to the point of fracture) and fracture toughness of bone most likely stem from age-dependent changes in the mineral and collagen phases of the bone. Structural changes in collagen molecules (denaturation or unwinding) are significantly correlated with the elastic modulus and failure energy or toughness of the bone. Meanwhile, bone mineral content has a significant correlation with the elastic modulus, yield strength and ultimate strength of the bone. From an engineering perspective, the mineral phase in the bone most likely imparts stiffness and strength to the bone, while the collagen network functions as a reinforcing phase to sustain its toughness.
When the normal triple helical structure of collagen molecules is unwound or cleaved, the molecules become less stretchable, potentially lose their strength and capacity to absorb energy and consequently make the bone more brittle.

8.3 Modelling

Mathematical modelling, especially with the improvements in computer architecture, has become an important tool for explaining and predicting complex biophysical phenomena. Bones may be modelled (including void spaces and interconnectivity) by hierarchical micromechanics, finite element (FE) micromechanical computations or the homogenisation theory. With more information being available on the structural organisation of collagen and HA at the molecular ultrastructural level, it will now be possible to develop better models.

8.3.1 Conventional Modelling

With state-of-the art modeling gaining importance, conventional modelling has taken a back seat. In micromechanical modelling, a conventional modeling technique, apart from the knowledge of the bulk properties of the trabecular bone, some idea about its elastic properties is also necessary. The elastic modulus may be determined mechanically (by machining and buckling), or by ultrasonography, acoustic microscopy and nanoindentation techniques. These nanoscale measurements (e.g., those obtained by atomic force microscopy (AFM)), help in analysing the mechanics of the smallest structures in the bone.

Firstly, the elastic anisotropy (Ac* for compressive and As* for shear) percentages have to be calculated. In the Haversian bones, both these values are considerably lower than that in plexiform ones.

$$\text{AC*}(\%) = 100\frac{K^V - K_R}{K^V + K_R} \tag{8.7}$$

$$\text{As*}(\%) = 100\frac{G^V - G_R}{G^V + G_R} \tag{8.8}$$

where K^V and K_R are Voigt (uniform strain across an interface) and Reuss (uniform stress across an interface) bulk moduli, while G stands for the corresponding Voigt and Reuss shear moduli, respectively. For micromechanical modelling, a Haversian bone is usually taken as a hierarchical composite (hollow fibre composite). However, FEM (described in the next section) is gradually taking over the more conventional modeling techniques.

8.3.2 FEM

FEM (finite element method) is one of the methods used for modeling bones where the solution domain (of the partial differential equation, PDE) is approximated by a grid of uniformly spaced nodes. The solution domain can be discretised into a number of uniform or non-uniform FE connected via nodes. The change of dependent variable with location is approximated within each element by the interpolation function that is defined relative to the values of the variables at the nodes associated with each element. The original boundary value problem is then replaced with equivalent integral formulation. The interpolation functions are then substituted into an integral equation, integrated and combined with results from all other elements in solution domain. The results of this procedure can be reformulated into a matrix equation of form $Af = b$, which is subsequently solved for unknown variables. It may also be applied to a 3-D domain.

The formulation of the finite element approximation starts with the Galerkin approximation $(\sigma\nabla\phi, \nabla\bar{\phi}) = -(I_v, \bar{\phi})$ where $\bar{\phi}$ is the test function. To discretise the continuous domain problem, the solution domain is made $\Omega = \bigcup_{e=1}^{E} \Omega_e$ and a finite dimensional subspace $V_h \subset V = \{\bar{\phi} : \bar{\phi}$ is continuous on W, $\nabla\bar{\phi}$ is piece-wise continuous on $W\}$. For approximating a volume conductor in three dimensions, the 3-D domain W represents the geometry of the volume conductor and breaks it up into discrete elements to form a finite dimensional subspace W_h. For 3-D domains, the function may be denoted as either tetrahedra:

$$\overline{\Phi} = \alpha_1 + \alpha_2 x + \alpha_3 y + \alpha_4 z \tag{8.9}$$

or hexahedra

$$\overline{\Phi} = \alpha_1 + \alpha_2 x + \alpha_3 y + \alpha_4 z + \alpha_5 xy + \alpha_6 yz + \alpha_7 xz + \alpha_8 xyz \tag{8.10}$$

For further details about FEM modelling, please consult References [1], [6] and [7].

To cite an example, for the formulation and evaluation of a structural model for trabecular bone material (e.g., of the vertebra), the work will consist of several parts:

i Formulation of an analytical structural model for the regular hexagonal unit cell that relates nodal forces to nodal displacements (elastic and elastoplastic deformation regimes).
ii Determination of the elastic and incremental elastoplastic structural properties of the unit cell using the analytical model.
iii Validation of the (foam) model as an accurate model of a vertebral trabecular bone.

This, in turn, will involve several parts:

i Evaluation of the overall behaviour of a cylindrical bone specimen subjected to uniform compression in a 'virtual' finite element simulation.
ii Generation of an identical prototype using a stereolithographic prototyping machine (e.g., the one performed by the Center for Ceramic Research, Rutgers University) which will be tested in a mechanical testing machine.
iii Comparison of the actual results with the computer generated results which will establish the validity and accuracy of the model.
iv Finally, comparison with experimental results obtained for actual vertebral trabecular bone specimens which will establish the hexagonal open-celled foam as an accurate representative.
v Implementation of the foam model of a trabecular bone into a finite element analysis (FEA) of loading applied to a whole vertebra.

This will help to develop a versatile, numerically efficient and powerful model for trabecular bone that accurately represents the microstructure of the material. The model will be useful in the following areas:

i Study of damage accumulation in the trabecular structure with loading
ii Computer aided design, development, and evaluation of orthopedic implants using finite element analysis
iii Finite element crash/impact load simulations
iv Investigation of the effects of osteoporosis and other clinical problems (like burst fracture) on the structural integrity of the vertebra
v Analysis of the effects of structural orientation and relative density on the overall behaviour of trabecular bone

8.4 Further Reading

[1] Van de Velde, E. F. *Concurrent Scientific Computing.* Springer-Verlag. 1994.

[2] Keaveny, T. M. and Hayes, W. C. A 20-year perspective on the mechanical properties of trabecular bone. *J. Biomech. Eng.* 115: 535. 1993

[3] Hogan, H. A. Micromechanics modeling of haversian cortical bone properties. *J. Biomech.* 25: 549. 1992

[4] Katz, J. L. and Meunier, A. A generalized method for characterizing elastic anisotropy in solid living tissues. *J. Mat. Sci. Mater. Med.* 1: 1. 1990.

[5] Hoole, S. R. H. Computer-aided analysis and design of electromagnetic devices. NY: Elesevier. 1989.

[6] Burnett, D. S. *Finite Element Method.* Reading, Mass., Addison-Wesley. 1988.

[7] Akin, J. E. *Finite Element Analysis for Undergraduates.* NY: Academic Press. 1986.

[8] Currey, J. *The Mechanical Adaptation of Bones.* NJ: Princeton Univ. Press. 1984.

[9] Hastings, G. W. and Ducheyne, P., Eds. *Natural and Living Tissues.* Boca Raton, FL: CRC Press. 1984.

[10] www.caip.rutgers.edu/~langrana

8.5 Questions

1. Descriptive Essay Questions

1. How do the models for various types of bones differ?
2. Describe finite element modelling of the bone.

2. Write Short Notes on

1. Haversian bone
2. Viscoelasticity
3. Compressive anisotropy
4. Shear anisotropy
5. Grid of uniformly spaced nodes

3. Multiple Choice Questions

Choose the single best response.

1. H (hydrogen) bonds are stronger than which of the following bonds ?
   (A) van der Waal's
   (B) metallic
   (C) covalent
   (D) ionic.

2. Covalent and ionic solids are not suitable as structural components since
   (A) they are ductile
   (B) they are brittle

(C) they have weak bonds
(D) they have high elastic moduli

3. Linear combination of Maxwell and Voigt-Kelvin element is a ___ model
(A) 1-parameter
(B) 2-parameter
(C) 3-parameter
(D) 4-parameter

4. While modelling the bone, finer details of structural mechanics may be determined by
(A) electron microscopy
(B) light microscopy
(C) atomic force microscopy
(D) magnetic resonance microscopy

5. A model of a bone may help in
(A) CAD
(B) understanding the causes of fracture
(C) predicting maximum load to be borne
(D) all of the above

Answers

1. A; 2. B; 3. D; 4. C; 5. D

9

Biomechanics of Implant Tissues

9.1 Introduction

Biomechanics (BME), as mentioned before, is the science of the forces acting internally or externally on the living organism or the cell. It may also be defined as the application of mechanical laws to living structures, as to a locomotor system. A rapidly growing branch of BME is *tissue engineering*, defined as the application of technical principles to the design, construction, modification, growth and maintenance of living tissues. A *tissue* is nothing but a conglomeration of (biological) cells, organised to carry out a specific function. Tissue engineering has gained in scope and importance due to the prolific development in cellular and molecular biology, as well as in materials, chemical and mechanical engineering. We can safely say that the manipulation and reconstitution of tissue functionis going to be a key player in cell and gene therapies and also in transplantation medicine.

Properties of materials like the biomolecular properties of a surface and protein adsorption may be discussed both in molecular and cellular scales. These properties especially the physicochemical rate processes affect the tissues in many ways. Fluid mechanical forces and mass transfer rates influence important attributes of tissue function and relate to the physical constraints under which tissues operate. *In vivo* cellular microenvironment is critical for the functioning of any tissue. It is important to keep in mind the interactions between specialised cells and their adnexal/accessory cells while reconstituting tissue function *ex vivo*.

Tissue engineering includes both *in vitro* construction of bioartificial tissues from cells isolated by enzymatic dissociation of donor tissue and *in vivo* alteration of cell growth and function. The application may be an implantable device (like vascular grafts), an extra corporeal device (like bone or cartilage implants), cell production (e.g., bioartificial skin) and *in situ* tissue growth and repair (e.g., bioartificial pancreatic islets, neurotransmitter secreting cells, bioartificial liver, *in vitro* hematopoiesis, nerve regeneration and artificial skin).

9.2 Molecular Properties

The biomolecular properties of a cell surface affect the cellular physiological processes like adhesion, spreading, activation, recruitment, migration, proliferation and differentiation. If we can understand these cell surface interactions properly, we may be able to design substances (e.g., drugs) to enhance or reduce cell–substrate interactions. Increasing cell adhesive properties may help in tissue regeneration or biomaterial integration, while inhibiting it may help in preventing metastasis or fibrosis.

Appropriate adhesion of cells to a culture is necessary in many areas of tissue engineering, like, biohybrid skin dressings, synthetic articular cartilage, hepatocyte scaffolds.

For studying the cell surface interactions, various simple models simulating the basement membranes, cell surfaces or extracellular matrices have been developed. These models immobilise the biologically active ligands (both natural and synthetic) onto different substrates to produced bioactive surfaces which can be defined chemically. To understand the molecular mechanisms of cell-mediated bioadhesion, the ligand-functionalised 2-D substrates are prepared analogous to labelling ligands, immunomatrices and immobilising enzymes.

Multiple factors govern bioactivity, while immobilising ligands for cell-adhesive substrates. The reactive sites on the ligand molecule must be sterically accessible, cell-receptor interactions must not be bound by physical constraints like burial of active site or partial denaturation of the ligand by immobilisation. If spacer arms are included on the surface, these problems may be solved.

Physicochemical absorption of ligands onto polymeric substances through covalent immobilisation precludes desorption. However, this process may call for activation. During activation care has to be taken to avoid homopolymerisation or cross-linking.

Since biological ligands are usually immobilised for these studies, water is often the best coupling medium.

Cell surface studies usually involve chemical groups like hydroxyls, amines, carboxylic acids and thiols for immobilising the ligands by surface bioconjugation.

The surface containing reactive groups for ligand bioconjugation can be prepared with wet chemical methods or plasma deposition of polymeric substances.

Hybrid copolymers can be synthesised with a biologically active ligand. They can then be coated onto a substrate or cross-linked to a 3-D network. Bioactive substrates may then be produced without further ligand immobilisation as the ligand is a component of the copolymer.

For measuring the concentration of ligands immobilised onto a surface, radiolabelling, photochrome labelling, surface analysis or gravimetry methods may be applied. Rigid controls must be maintained to distinguish the specific bioconjugation (covalent immobilisation) and non-specific (physicochemical) adsorption by various bioactive ligands like proteins.

Immobilised ligands may be used for appreciating the different molecular facets of cell adhesion. Extracellular matrix proteins like fibronectin, laminin, vitronectin and collagen, as well as adhesion molecules like ICAM-1, VCAM-1, PCAM-1, and sialyl Lewis X, interact with cell surface receptors to mediate cell adhesion.

A ubiquitous signal present in most cell adhesion proteins is the tripeptide sequence of RGD (Arg-Gly-Asp or Arginine-Glycine-Aspartate), which interacts with the integrin family of cell surface receptors. Therefore, synthetic RGD-grafted substrates have been utilised in many areas of tissue engineering. This certainly enhances the chance of the biocompatibility of the implant where this sequence is artificially introduced through various methods.

9.3 Cellular Properties

Earlier studies have confirmed the roles for adsorbed proteins in subsequent cell protein surface interactions on a molecular level. The peptide sequences (e.g., RGD) binding to the cellular receptors integrins (Section 9.2), have been identified, characterised, and synthesised. Now, selective or enhanced *in vitro* cell cultures and selective *in vivo* cellular responses like endothelialisation is possible even without inflammation, infection or thrombosis.

However, first we must have a clear idea about the basics of protein adsorption. Some key concepts for understanding this are structure and heterogeneity of the proteins, factors affecting the thermodynamics and kinetics of adsorption and its reversibility, and also the dynamics of multi-component adsorption. The amino acids of a peptide, polypeptide or a protein are known as *residues*. The higher order protein structure that describes the spatial relationship between the component amino acids of the protein is called the *conformation*. The particular region of a protein where a cell or an antibody can bind is the *epitope*. The organisation defines the manner how a protein resides upon a surface (i.e., the existence, arrangement and availability of various epitopes).

Incidentally, plasma concentration denotes the volume fraction of blood plasma in the adsorption medium when protein adsorption from different dilutions of plasma is measured. The changes in protein conformation and organisation occurring due to adsorbed proteins residing onto a surface are called *post-adsorptive transitions*. The amount of radioactivity detected per unit mass of a specific protein is known as the *specific activity*.

The *Vroman effect*, an important phenomenon, is the collective term describing:

(i) maximal and speedy adsorption of a specific protein from a multi-component medium.
(ii) maximal equilibrium adsorption from intermediate dilution, and,
(iii) reduction of the plasma concentration, at which adsorption is maximum, with enhanced adsorption time.

The *Vroman peak* is the adsorption maximum in the adsorption versus log (plasma concentration) curve when protein adsorption from different dilutions of plasma is measured.

The biomaterials processed for tissue engineering are in mm or nm range. Immunoprotective devices have semipermeable membranes which prevent host immune cells (antibodies) from entering. In the open type membranes, pores of size greater than 10 μm facilitates free transport of cells and molecules between host tissues and transplanted ones. Most devices are synthetic rather than natural since common organic polymers can be produced easily, inexpensively. It is also simpler to standardise them. Extracellular matrix (ECM) molecules help in cell adhesion and organisation. An ECM analog is a model of ECM comprising a highly porous graft of copolymer of collagen (Section 7.3) and a glycosaminoglycan.

Unlike drugs, which replenish a molecular missing function, regeneration templates rectify the missing function at the level of tissue or organ. For instance, a dermis regeneration template (DRT) is a graft copolymer of type I collagen and chondroitin-6-sulfate. It has an average pore diameter of 20–125 μm, degrading *in vivo* to about 50% in a fortnight and inducing partial regeneration of the dermis in wounds from which the dermis has been fully excised. If keratinocytes are seeded before grafting, both dermis and epidermis synthesis can be induced *in vivo*.

9.4 Physicochemical Rate Processes

Vascular cells are always exposed to fluid mechanical (hemodynamic) forces (Section 14.3.1) due to blood flow. The forces generated there include frictional force (fluid shear stress, Section 9.3) caused by the tangential blood flow across the endothelium, tensile stress due to circumferential blood vessel wall deformations, and a net normal stress because of a hydrodynamic pressure differential across the vessel wall.

Cell function changes induced by mechanical force have been studied *in vitro* with cultured monolayers of bovine aortic endothelial cells (BAECs) and human umbilical vein endothelial cells (HUVECs). Shear stress is usually generated *in vitro* by flowing fluid across endothelial cell monolayers under controlled kinematic conditions, commonly in the laminar flow regime. Most of the times either parallel plate or cone-and-plate geometries are used. Reference or physiological levels of venous and arterial shear stress range between 1–5 dynes/cm^2 and 6–40 dynes/cm^2, respectively.

Shear stress mediated cell endothelium interactions usually affect the cell morphology, cytoskeletal rearrangement, signal transduction and mass transfer (e.g., due to ATP-mediated rise in calcium ions), metabolite secretion (e.g., arterial levels of shear stress enhance the secretion of tissue plasminogen activator or tPA) and also gene regulation. The nucleotide sequence of the SSRE (shear stress response element), GAGACC has been identified in genes responsive to shear stress.

Mass transfer purveys the major constraints on the anatomy and biochemistry of all the structures of the living organisms. The mathematical bases for quantification of these properties are: self similarities and cross-species correlations, time constants, Brownian motion and diffusion along concentration gradients in (often) multi-component solutions.

9.5 *In vitro* Studies

Most eukaryotes, and especially all mammals, begin the journey of their life from a single totipotent stem cell termed as zygote (Up to the first 2 weeks it is actually the fertilised ovum, up to 2 months or 9 weeks it is known as the embryo, and after that up to delivery, it is called the foetus. However, the embryo itself starts differentiation). In 1981, two scientists at Cambridge (Evans and Kaufman), isolated embryonic stem (ES) cells from a mouse embryo at the blastocyst stage, and cloned them *in vitro* for many generations, maintaining their totipotency. This has been followed by 'gene knockout' – mutating cellular DNA *in vitro*. Often the term '*ex vivo*' is used interchangeably with *in vitro*.

9.6 *In vivo* Studies

In vivo interactions like cell-cell interactions, cell-matrix (ECM) interactions and cellular interactions with the fluid phase (containing growth factors) of the tissue, cause cell translocations, as well as various changes in the events of the cell cycle. Motile interactions are caused by the different feedback loops between the complex dynamic systems comprising fluid space, matrix and the other cells in the tissue.

Biohybrid systems are more useful for tissue engineering (especially for kidney cells) rather than only synthetic materials. In the case of skeletal muscle engineering, myoblast transfer and gene therapy has been used for treating genetic disorders like Duchenne muscular dystrophy (DMD) or pseudo-hypertrophy muscular dystrophy (where the gene for the dystrophin protein is absent or reduced). For gene therapy, direct injection of plasmid DNA containing gene for dystrophin may be used or it may be mediated either through a retrovirus or adenovirus.

9.7 *In situ* Studies

Bone marrow transplant (BMT) is usually not placed in its original site (it is injected into the bloodstream, from where it reaches its destination), but most of the *ex vivo* or *in vitro* engineered tissues like liver, nerve cells, muscle cells, cartilages and kidney cells, like prostheses are actually operated into their original or nearby positions. This is also known as *in situ* placement.

Hydrogels are cross-linked hydrophilic polymers formed *in situ*. They can be formed easily, swell up with water and give a soft tissue effect. They have high permeability and a good biocompatibility. Visible or ultraviolet light can be made to interact with light sensitive substances called *photoinitiators*, which create free radicals for initiating polymerisation to form cross-linked hydrogels (common ones include PEG or polyethylene glycol). Photoactive PVA (polyvinyl alcohol) hydrogels have been used as scaffold in tissue engineering. Cells seeded within the PVA culture can form ECM.

Non-adhesive PEG hydrogels are now being modified with RGD peptide sequences to facilitate adhesion, spreading and cytoskeletal organisation of osteoblasts (bone forming cells).

9.8 Further Reading

[1] Morgan, J. R. and Yarmush, M, L, Eds. *Tissue Engineering Methods and Protocols*. Totowa, NJ: Humana Press. 1999.

[2] Nguyen, K. T. and West, J. L. Photopolymerizable hydrogels for tissue engineering applications. *Biomaterials* 23: 4307–4314. 2002.

[3] Schmedlen, R. H, Masters, K. S. and West, J. L., Photocrosslinkable polyvinyl alcohol hydrogels that can be modified with cell adhesion peptides for use in tissue engineering. *Biomaterials* 23: 4325–4332. 2002.

[4] Burdick, J. A. and Anseth, K.S. Photoencapsulation of osteoblasts in injectable RGD-modified PEG hydrogels for bone tissue engineering. *Biomaterials* 23: 4315–4323. 2002.

[5] Stokke, B. and Elgsaeter, T, Eds. *Chemical and Physical Network Formation and Control of Properties*. West Sussex: Wiley. 1999.

9.9 Questions

1. Descriptive Essay Questions

1. What do you understand by tissue engineering ? Give some examples.
2. How do physicochemical rate processes affect tissues ?

2. Write Short Notes on

1. Biomolecular properties of surfaces
2. Protein adsorption

3. *In vivo* cellular microenvironment
4. Bioartificial tissue
5. *In situ* tissue growth and repair

3. Multiple Choice Questions

Choose the single best response.

1. The osmolality of plasma in mosm/L (milliosmolal) is
   (A) 250
   (B) 270
   (C) 290
   (D) 310

2. Vroman effect includes all except
   (A) maximal and speedy adsorption of a specific protein from a multi-component medium
   (B) maximal equilibrium adsorption from intermediate dilution
   (C) reduction of the plasma concentration, at which adsorption is maximum, with enhanced adsorption time
   (D) none

3. *Ex vivo* generation of human tissues is
   (A) difficult
   (B) due to regenerative or hyperplastic capacity of them
   (C) most successful in the case of bone marrow
   (D) all

4. All except which form ECM ?
   (A) erythropoietin
   (B) fibronectin
   (C) laminin
   (D) vitronectin

5. Tissue engineering in the nervous system may be used for all except
   (A) functional replacement of a missing neuroactive component
   (B) rescue or regeneration of damaged central nervous system neuron
   (C) rescue or regeneration of damaged peripheral nervous system neuron
   (D) neural coupling elements for man-machine interfaces

Answers

1. C; 2. D; 3. D; 4. A; 5. B

10

Sports Biomechanics

10.1 Introduction

Some knowledge of exercise physiology helps the biomedical engineer to properly design devices for use with or by humans or animals. The engineer can also borrow ideas from this physiology and apply them to other conditions.

While the metabolic rate of a body under fever conditions may be raised by 100%, in a marathon runner, it may exceed 2000%. Muscle strength, pulmonary ventilation and cardiac output – all depending on muscle mass – in females is 65–75% of the values in males. However, females can achieve the same maximal force of muscle contraction as men, i.e., 3–4 kg/cm^2, although they have less muscle mass. The male hormone testosterone is strongly anabolic and causes increased protein deposition (especially in muscles) whereas the female hormone estrogen causes more fat deposition in breasts, hips and subcutaneous tissues. The *holding strength of a muscle* is the force required to stretch out a muscle after it has contracted. It is about 40% greater than the maximal contractile strength.

Endurance is dependent on the maintenance of nutritional supply to the muscles, which usually comes from

(i) the phosphagen system – ATP and creatine phosphate or phosphoryl creatine – which can supply energy for 8–10 seconds maximal contraction, like that required in a 100-m race,

(ii) the glycogen–lactic acid or anaerobic system which can give a spurt of energy for maximal muscle contraction of 1.3–1.6 minutes and

(iii) the aerobic system for recovery which converts lactic acid to pyruvic acid. Pyruvic acid is then oxidised or reconverted to glucose (in the liver).

10.2 Muscle Strength

As mentioned in Section 10.1, skeletal muscle in human beings (like other mammalians) can exert 3–4 kg of tension per square centimetre of cross-sectional area. To cite two examples, one of the muscles of the triceps surae (calf muscles) or gastrocnemius has a relatively large cross-sectional area and, thereby, through its large tension can not only support the body weight during climbing, but can also resist a much greater force when a person runs or jumps. Another major muscle of the buttock, the gluteus maximus (does the Asterix comics characters' names come to mind?) can exert a tension of about 1200 kg. The total tension that may be developed by all the muscles of an adult man can be nearly 22 tons.

10.3 Muscle Power

The *power* of a muscle is the amount of work performed per unit time. Muscle power is usually graded into six levels, from zero to five. The zero level denotes absence of any contraction. Grade 1 denotes a muscle flicker. Grade 2 means gravity eliminated movement (i.e., the subject can move a limb, if it is placed on a bed or table). Grade 3 is when the subject can resist the force of gravity (e.g., standing without support) but cannot resist any force applied by an external source. In the case of Grade 4 muscle power, the subject is capable of resisting mild force (in the opposite direction) from the external source, though it is below normal. Grade 5 is accepted as normal muscle power clinically.

Before proceeding further, let us have a brief discussion about how skeletal muscles contract.

Usually when a muscle contracts, the contractile elements (thin filaments comprising actin, troponin and tropomyosin) shorten by sliding over the thick filaments (consisting of myosin). This phenomenon is also known as the *sliding filament theory*. However, since the muscles have viscoelastic elements in series with the contractile mechanism, contraction is possible even without an appreciable reduction in the length of the whole muscle. This type of contraction is called *isometric* (equal magnitude or length) *contraction* as in the case of standing (the quadriceps femoris muscle tenses to tighten the knee joint). The most common type of contraction is found in most forms of exercise and is known as *isotonic* (equal tension) *contraction* because there is a constant load with shortening of the length of the muscle, e.g., weight lifting.

Funnily, since in an isometric contraction there is no displacement, no work (force ′ distance) is done either. Another interesting point is that a muscle may do negative work (lengthening instead of shortening), e.g., when a heavy object is lowered down. However, most contractions are of the mixed type. If we take running as an example, there is isometric contraction when the legs (soleus muscle) hit the ground and isotonic contraction when the limbs are moved (by the gastrocnemius muscle).

The relation between muscle length, tension and velocity of contraction is dealt with in Section 14.3.

10.4 Endurance

Endurance is dependent on the maintenance of the nutritional supply of muscles which will be discussed in the next section. To meet the enormous energy demands during exercise, most of the physiological responses have to be optimised. For instance, walking occurs naturally at a certain pace (Section 14.4) representing the smallest energy expenditure. Similarly, transition to running (from walking) occurs when less energy is spent in running. Endurance time, in accordance with the system of metabolism used, is summarised in the next section (Table 10.1)

Heat dissipation during exercise is also associated with higher resetting of the hypothalamic 'thermostat' centre. Since muscular activity is hardly 20–25% efficient (smaller muscles are even less efficacious), heat results from the remaining 75–80% energy supplied to the muscle. Only if the exercise is performed at a low rate, does the performance time become long enough to store significant amount of heat. Counter-current heat exchange in the arterio-venous system conserves some heat. Thermal stress induces blood to return through superficial veins (instead of deep ones), leading to an increase in skin temperature, followed by convection and radiation of heat to the atmosphere. Cutaneous vasodilation also increases surface heat loss, although it poses an extra burden on the heart as it has to deliver more blood to skin and blood vessels. Heart rate is also enhanced by heat and sweating starts, especially if the environment is humid. Excess sweating leads to hemoconcentration, increased blood viscosity and further workload on the already overactive heart. Incidentally, if the sweat does not evaporate, (absorbed by clothes, the atmosphere or falling to the ground), the cooling effect (670 W h/kg) is not achieved.

10.5 Nutrition and Metabolism of Muscles

As mentioned in the introduction, muscle energy is usually derived from (i) the phosphagen system, (ii) the glycogen–lactic acid or anaerobic system and (iii) the aerobic system. A comparative assessment of the three systems is presented in Table 10.1.

Table 10.1 Comparison between muscle energy systems

System	Maximal rate of production of ATP (M/min)	Endurance time	Muscle power (kg m/min)
Phosphagen	4	8–10 seconds	7000
Anaerobic	2.5	1.3–1.6 minutes	4000
Aerobic	1	Unlimited (as long as nutrients last)	1700 (up to about 30 minutes)

In the above table, unlimited endurance time through the aerobic processes actually depend on the type of diet taken by the subject. For instance, on a high carbohydrate diet, muscles may store 40 gm/kg of glycogen, while on a mixed diet (carbohydrate, protein and fat), the glycogen stored may be 20 gm/kg of muscle mass and on a high fat diet, the glycogen stored will amount to only 6 gm/kg. On such diets, the endurance time (at the speed of a marathon runner) will be respectively 4 hours, 2 hours and 1 hour 25 minutes. The type of energy system utilised will depend on the specific (athletic) activity. To cite concrete examples, for a 100-m sprint, only a phosphagen system may be sufficient, although for a 200-m dash, phosphagen, as well as an anaerobic system would be required. For a 400-m run more of anaerobic and a slight aerobic system has to be mobilised, for a 800-m run more of aerobic and less of anaerobic systems will be working, and for a marathon race (26.2 miles or 42.2 km), it is mostly the aerobic system which is essential.

Removal of one high energy phosphate (HEP) bond from ATP gives 7300 cal/mole of ATP energy release. In the case of CP, 10300 cal/mole of energy is released and the total store is about 2–4 times the ATP store. In an anaerobic system, 1mole glucose liberates 4 ATP, and it is produced 2.5 times faster than in oxidative metabolism.

Based on the metabolic activity, there are two types of skeletal muscle fibres. *Slow* or *oxidative* or *red* or *type I* muscle fibres are half in diameter compared to *fast* or *glycolytic* or *white* or *type II* muscle fibres. The glycolytic capacity and myosin isoenzyme ATPase rate is 2–3 times more in type II fibres. Type I fibres have a high oxidative capacity (more mitochondria, capillary density and myoglobin, which is the reddish pigment responsible for carrying oxygen to the muscles), while type II fibres have a much higher Ca^{++} pumping capacity of sarcoplasmic reticulum.

The resting skeletal muscle has a blood flow (BF) of about 2–4 mL/100 g/min. With muscular contraction (if greater than 10% of its maximal tension), the blood vessels within the muscle get compressed. When more than 70% of maximal tension is developed, BF is completely stopped. Nevertheless, in between contractions, BF per unit time in a rhythmically contracting muscle is increased up to thirty times.

If the muscles are contracted with at least 50% maximal force for a few times per day, thrice weekly, muscles may grow through *hypertrophy* – a phenomena where myofibrils increase in number, mitochondrial enzymes are raised by 120%, phosphagen energy system components are elevated by 60–80%, stored glycogen rises by 50% and stored triglycerides escalate by 75–100%.

Anaerobic threshold is defined as the transition between exercise levels that can be sustained through nearly complete aerobic metabolism and those that rely on at least partially anaerobic metabolism. Above it, the blood lactate increases and the relationship between ventilation and oxygen uptake becomes non-linear.

10.6 Oxygen Consumption

Exercise does not affect only the muscles but also the cardiovascular and respiratory systems. Athletic training increases the maximal oxygen consumption ($\dot{V}_{O_2 max}$) that exercise can induce in a person. The reference values of $\dot{V}_{O_2 max}$ are about 38 mL/kg/min in an active healthy man and about 29 mL/kg/min in an active healthy woman, although it is lower in sedentary people. $\dot{V}_{O_2 max}$ is the product of the maximal cardiac output and the maximal tissue extraction of O_2. Both these factors are raised by regular physical exercise. With age, this value gets reduced.

At maximal exercise, pulmonary ventilation is 100–110 L/min, but the maximal breathing capacity (MBC) is greater than 150 L/min. This built-in reserve is helpful while exercising in high altitude, high temperature and also in some respiratory abnormalities. A trained athlete has a pulmonary O_2 diffusion capacity almost four times the normal resting value in non-athletes (< 23 mL/min/mm Hg). Athletic training also raises stroke volume and reduces heart rate. High temperature can be lethal to athletes as it causes decreased loss of heat and dehydration.

Oxygen debt or *deficit* is the accumulated difference between actual oxygen consumption at the beginning of the exercise and the rate of oxygen consumption that would exist if oxygen consumption rose immediately to its steady-state level corresponding to the exercise level.

The difference between resting oxygen consumption and the accumulated rate of oxygen consumption after the completion of exercise is known as the *excess post-exercise oxygen consumption (EPOC)*.

10.7 Performance Engineering

Ergonomics is the discipline, which investigates and applies information about human requirements, characteristics, abilities and limitations to design, development and testing of equipment, systems and jobs. It attempts to improve human performance by optimising relations between people and their work environment. Also see Chapter 14.

Performance capacities are the dimensional capabilities or resources like speed, accuracy, strength, that a given human system like the visual, auditory, speech or control system of gait and posture, possesses to perform a given task. This capacity may be ascertained by *performance assessment* - the process of determining the level or degree up to which a system can perform a specific task, given the required demands and resources' availability of the system involved. The *performance resource* is nothing but the functional unit and an associated dimension of performance (e.g., strength of knee extensors) available to a system. Therefore, *human performance engineering* is a dynamic area with quick overall changes and constantly developing new techniques affecting the possibilities of application. The more important aspects of it are related to human performance modelling, measurements and data-related topics.

The common models encountered in human performance are conceptual, statistical based predictive, causal predictive, and data models. The statistically based predictive models are usually regression models. Various simulation software packages are also available for modelling the biomechanical, sensory, cognitive and structural details. *Reductionism* is a methodology for understanding the present tense, rather than predicting the future potential of a model. Unlike biological systems, in artificial ones, reverse engineering (the reductionist approach) has been successful in creating functional equivalent replicas. An important distinctive feature in all living beings is individual variations. Their standardization and proper measurement require newer and more sophisticated devices. The best example of individual variation, related to sports activity is that trained athletes have a larger stroke volume and lower heart rate, compared to untrained people, both during exercise, as well as during rest. Usually they have larger hearts too.

In a nutshell, a rather high degree of sophistication is essential to bridge the gap between understanding the science of human performance and engineering tools to enhance and replace it. This is certainly an area where multi-disciplinary interaction is of utmost necessity. The branches of occupational therapy, occupational rehabilitation and ergonomics, all depend on the outcome of human performance engineering.

10.8 Further Reading

[1] Harris, S. S., Caspersen, C. J., DeFriese, G. H. and Estes Jr., E. H. Physical activity counseling for healthy adults as a primary preventive intervention in the clinical setting - Report for the US Preventive Services Task Force. *JAMA* 261: 3590–3598. 1989.

[2] Hays, K. F. The use of exercise in psychotherapy. pp.155–168 in *Innovations in Clinical Practice*, Vol. 12, eds. van de Creek, L., Knapp, S. and Jackson, T.L. Sarasota, FL, USA: Professional Resource Press. 1993.

[3] Morgan, W.P. Physical activity, fitness and depression. pp. 851–867 in *Physical Activity, Fitness and Health : International Proceedings and Consensus Statement*, eds. Bouchard, C., Shephard, R.J. and Stephens, T. Champaign IL: Human Kinetics Publisher. 1994.

[4] Johnson, A. T. *Biomechanics and Exercise Physiology*. John Wiley & Sons. 1991.

[5] Astrand, P.O. and Rodahl, K. *Textbook of Work Physiology*. McGraw Hill. 1970.

10.9 Questions

1. Descriptive questions

1. Describe muscle metabolism.
2. How can athletic training improve endurance and performance?

2. Write Short Notes on

1. Muscle power
2. Endurance
3. Oxygen consumption
4. Respiratory quotient
5. Myoglobin

3. Multiple Choice Questions

Choose the single best response.

1. Respiratory quotient (RQ) of fat, protein, carbohydrate respectively are:
   (A) 0.7, 0.8, 1.0
   (B) 0.8, 0.7, 1.0
   (C) 0.8, 1.0, 0.7
   (D) 1.0, 0.8, 0.7
2. Ergonomics deals with
   (A) human requirements, characteristics, abilities and limitations
   (B) design, development and testing of equipment, systems and jobs
   (C) optimisation of relations between people and their work environment
   (D) all.
3. For a 100-m swimming match, which system of energy utilisation is essential ?
   (A) more of aerobic and less of anaerobic systems
   (B) more of anaerobic and a slight aerobic system
   (C) mostly the aerobic system
   (D) mostly the phosphagen system
4. $V_{O_2\,max}$ is dependent on
   (A) age
   (B) sex
   (C) habits
   (D) all
5. Performance engineering deals with
   (A) artificial joints
   (B) artificial limbs
   (C) modelling of human performance and its measurements
   (D) none of the above.

Answers

1. A; 2. D; 3. B; 4. D; 5. C

Section 4

Biomaterials

A *biomaterial* is any substance (other than a drug), synthetic or natural, that can be used as a system or part of a system that treats, augments, or replaces any tissue, organ or function of the body.

Structure and properties of biological materials like bone, teeth and other connective tissues are discussed in Chapter 11.

Properties, compatibility characteristics and performance of implants are dealt with in Chapter 12. Chapter 13 discusses the structural properties of synthetic biomaterials.

11

Structure and Properties of Some Biological Materials

11.1 Introduction

15–16% of our body comprises of spaces between cells or *interstitium*. The fluid here is called interstitial fluid, ISF (ISF + blood plasma = extracellular fluid or ECF; the rest is all intracellular fluid or ICF). The solid structure (*matrix* of connective tissue) in the interstitium is mainly composed of collagen fibre bundles and proteoglycan filaments (chiefly hyaluronic acid).

So, our body is abundant in connective tissue (as you can see from Table 11.1), right from the blood to the bones.

The mechanical behaviour of soft biological tissues is characterised by pronounced non-linearities and viscoelasticity as well as by a multiphasic composition (fluid and solid). For mathematical modelling purposes, the finite element (FE) method (Section 4.3.2) is the method of choice in particular as biological structures usually exhibit an irregular geometry and often undergo large deformations. Thanks to the ongoing dramatic increase in computing power the mathematical modelling of organs and of the entire human body has become more and more close to real life.

The discipline of cardiodynamics (Section 14.3) is devoted to the analysis of the time-dependent deformation process of the beating heart from a mechanics' point of view. For this purpose, the heart is modelled as a structure exhibiting a two-fold anisotropy. First, a global anisotropy is derived from the gross rope-like architecture of the ventricles while a second type of anisotropy reflects the internal structure of the tissue. An unwinding of the myocardium along a virtual (anatomically non-existent) cleavage surface provided by the rope-like structure into more or less flat band lends itself to be applied for practical modelling purposes because realistic fibre patterns can be implemented in such a fashion at a reasonable expense, a task, which otherwise, i.e., in an intact spatial heart model would be difficult to achieve. Particular modelling efforts are directed towards more detailed geometries, especially for selected pathologic cases. Furthermore, it is directed towards the adaptation of fibre architectures which are in agreement with the MR measurements, the analysis of the active contraction phase of the heart cycle and the implementation of a bi-phasic constitutive relation to describe myocardial tissue.

Injury continues to represent a major public health problem, so much so that injury prevention remains an important task of trauma biomechanics. To this end, the analysis of injury mechanisms and the development of protection strategies are necessary. Because experiments under physiological conditions are not possible, the approach to be taken consists of the development of new computer models or a reformulation or extension of existing models, respectively, whereby new techniques, in particular finite element (FE, Section 4.3.2) techniques are applied. The ultimate goal is the development of an FE model of the entire human body.

A particular problem of rapidly increasing importance is associated with the cervical spine, in that low speed vehicle impacts, in particular from behind, are being more and more recognised as the cause of long-lasting disability. Such sequelae are observed even in impacts which were until recently believed to lead to loading levels on the occupant way below accepted tolerance limits.

Minute fluid displacements occurring in the cellular network during impact are hypothesised to cause impairment to neurological functions through the action of local shear stresses. Today, there exists insufficient knowledge with respect to the primary injury mechanisms involved as well as to the processes leading secondarily to long- lasting disease. Improved protection of the spine can be achieved on the basis of a better understanding of the mechanical processes of traumatic loading in typical accident situations.

In living tissues, substances involved in cellular activity, in particular water, electrolytes, nutrients and metabolites, are transported by systemic circulation, interstitial diffusion and the lymphatic system. Within soft tissues and organs, these mechanisms suffice in general to insure an adequate supply of substrates required for metabolic activity, they also provide a means for waste removal. However, the hard tissues of the skeletal system are relatively impermeable in comparison to soft tissues. Within the cortical bone, osteocytes depend on the transport pathways comprising the lacunocanalicular system for maintenance of metabolic traffic and interchange. Although studies have shown that diffusion is a major contributing mechanism for the transport of molecules and ions across the endothelial layer of bone capillaries, these diffusive mechanisms may in fact not be sufficient to insure adequate transport to and from osteocytes which are not in the close vicinity of the vascular canals. Given the lack of a true lymphatic system within skeletal tissues, some additional mechanisms of transport must prevail.

11.2 Connective Tissue

Based on the composition, connective tissue may be divided into six broad groups.

Table 11.1 Types of connective tissues

Cells in type of matrix	Examples
1. Free floating in fluid (plasma)	Blood
2. Semi-solid (jelly like) matrix with fine fibrils	Mesenchyme
3. Sem-isolid (jelly like) matrix with a fine network of extracellular reticular fibres	Reticular
4. Semi-solid (jelly like) matrix with thicker collagen and elastic fibre	1. loose fibrous (fibroblasts) 2. dense irregular (fibrocytes) 3. dense regular (fibroblasts between dense collagen bundles) 4. elastic 5. adipose (fat)
5. Solid elastic matrix with fibres	1. hyaline cartilage 2. white fibrocartilage 3. elastic/yellow fibrocartilage
6. Solid rigid matrix	1. bone 2. teeth

In the fifth type (cartilage), the first group (hyaline) is firm, flexible and forms the embryonic skeleton, costal cartilage (in ribs or chest cage), rings of trachea and the articular cartilages at the ends of bones. The white fibrocartilage is tough, resistant to stretching, slightly compressible, a great shock absorber and forms the intervertebral disks and symphysis pubis. The elastic or yellow fibrocartilage is flexible, resilient and forms the larynx, pinna (external ear) and epiglottis.

11.3 Collagen

Collagen purveys maximum tensile strength to the connective tissues, while thin preteoglycan filaments form fine reticular 'brush pile'. Connective tissue is so called as it binds epithelia, muscles and nerve cells together. There is no basement membrane in connective tissues and because of abundant intercellular ground substance, the cells are widely scattered. The cells may be local or histogenous (fibroblasts, histiocytes or clasmatocytes, fat cells, plasma cells, mast cells, pigment cells and mesenchymal cells) and migrated or hematogenous (lymphocytes, neutrophils and red blood cells, which may increase with inflammation). The matrix may be composed of reticular fibres and elastic fibres and keratin, apart from collagen or white fibres and proteoglycan.

For biological or tissue derived (isolated and purified) collagen please see Section 13.5.

11.4 Hydroxyapatite

Bone crystals vary between 3–7 nm in width and 20 nm in length, made up mostly of hydroxyapatites (HA) with the general formula ${Ca^{2+}}_{10}{}^{-x}(H_3O^+)_{2x}.\ ({PO_4}^{3-})(OH^-)_2$ or a simpler $Ca_{10}(PO_4)_6(OH)_2$. They may be used to reinforce metallic prostheses (e.g., artificial teeth or joints) for better bonding and stability (Section 13.5). Please also see Section 11.6.

11.5 Bone

Bone is a special form of connective tissue made up of microscopic crystals of phosphates of calcium embedded within a matrix of collagen (a complex 3-D organisation). Bones protect vital organs, permit locomotion and support loads against gravity. The protein in the bone matrix is mostly of the type I collagen that is also the major structural protein in tendons, skin and teeth. This protein (weight for weight as strong as steel) is made up of a triple helix of three polypeptides bound tightly together. Two of these are identical α_1 polypeptides encoded by one gene, and one is an α_2 polypeptide encoded by another gene. Collagens are a family of structurally related proteins maintaining the integrity of various organs. Up till now 15 different types have been identified, and, these are encoded by more than 20 different genes.

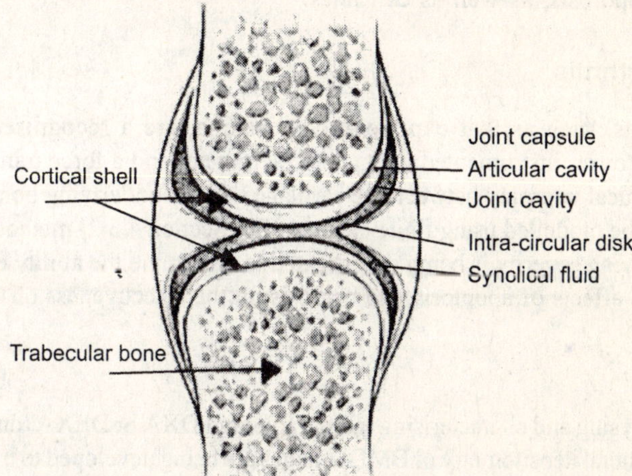

Fig.11.1 Structure of bone

Sodium, small amounts of magnesium and carbonate are also present in the bone. Most bones have an outer layer of compact bone surrounding the trabecular bone (Fig 11.1), and often a marrow cavity. The trabecular or spongy bone is made up of bone spicules separated by spaces. The compact bone is much denser and metabolically less active. In spongy bones, nutrients (about 25% of them) diffuse from the bone ECF into the trabeculae, while in compact bones, nutrients come through the Haversian canals, which contain blood vessels. Collagen is arranged concentrically around each Haversian canal, forming cylinders called *osteons* or Haversian systems. The viscoelastic properties of bone have already been described in Section 8.2.

The mechanical function of cartilage and bone tissue provides an alternate mechanism for promoting molecular transport, i.e., via load-induced fluid flow. Although the phenomena has not been fully elucidated yet, it has been shown that the mechanical loading of cartilage and bone tissue, associated with daily activities, results in the movement of the fluid contained within the tissue much like the loading of a stiff, water-filled sponge. So, it can be assumed that this enhances transport via convection. In addition to its hypothesised role in supporting cellular metabolism by regulation of local equilibria within the bone and cartilage, it has furthermore been shown that the stresses induced by such flow mediate biological remodelling processes intravasally (endothelial cells) as well as interstitially (osteocytes, chondrocytes). In the case of endothelial cells and osteocytes, it is postulated that shear stresses along the cell surface provide a mechano-transductive signal to trigger remodelling. Mathematical models on the basis of Biot's poroelastic theory, in combination with *in vitro* and newly developed *ex vivo* models are being applied in order to elucidate the importance of deformation-induced fluid flow in the bone.

Primary osteoarthritis

Significant changes in the subchondral bone of patients with primary osteoarthritis (osteoarthrosis) of the hip have been identified. Currently the nature of these changes is being investigated in great detail using a variety of physical methods. A new hypothesis for generalised osteoarthritis is that it is a systemic, metabolic disorder. This explains the involvement of multiple joints and the strong link with obesity. The biological origin of these changes is being studied using cultured cells from patients with osteoarthritis or osteoporosis, as well as cell lines.

Secondary osteoarthritis

Impact loads on joints, such as that experienced in trauma, are a recognised cause of secondary osteoarthritis. A drop tower, instrumented with an accelerometer and a force transducer may be used to investigate the mechanical interplay between the cartilage and the underlying bone during such loading. This system may also be modelled using finite element (FE, Section 4.3.2) methods. The response of the cells to impacts of varying severity is being investigated to determine the ability of the tissue to start the repair process and the affects of apoptosis and necrosis on the effectiveness of that process.

Osteoporosis

Novel methods of analysing and characterising radiographs and DXA or DEXA (dual energy X-ray beams for measuring bone mineral densitometry or BMD) images are being developed to try to increase our ability to measure changes in bone quality. There appear to be changes in the shape of the proximal femur which are strongly related to osteoporotic fractures of the femoral neck and identification of these is proving almost as good as bone mineral density measurements for predicting fracture. Ways are being developed for assessing the trabecular structure in bones and monitoring changes over time. Professor Stuart Ralston (University of Aberdeen, UK) has identified an alteration in the promoter region of one of the genes coding for collagen which could predispose a person to osteoporotic fractures.

Mechano-transduction and tissue engineering

Various devices for applying mechanical stimuli to cells *in vitro* have been built and are being used to study the responses of osteoblasts and chondrocytes. Connective tissues rely on mechanical stimuli to maintain a functional tissue matrix but still relatively little is known of how these signals are detected and transduced into synthesis of matrix molecules. Modelling of tissues as fibre composite materials shows how tapered fibres, which is how collagen fibres appear in many tissues, are advantageous for maximising stress transfer while minimising the metabolic cost of the materials.

Spinal biomechanics

In the area of spinal biomechanics, two novel models have been proposed to provide a better explanation of the load-bearing function of the spine. The first considers the spine to function as an arch-like structure and shows that in this way the closeness of the muscles to the spinal axis ensures stability while minimising the axial and shear stresses. The other shows how by controlling the axial stiffness of the spine it can be considered as an Euler pendulum (an inverted form of pendulum where the mass is above the support) thus maintaining stability while enabling precise control of movement.

11.6 Tooth

A tooth (Fig.11.2) is divided into the crown (projection above the gum margin), the root (fitting into sockets in the underlying bone) and neck (junction between the crown and the root, covered by the gum margin). The central cavity of hard portion or pulp (gelatinous matrix containing blood vessels and nerve fibres) is covered by dentine and cementum. *Enamel* is the hardest superficial portion covering the hard surface. Cementum is structurally similar to bone. Dentine contains 10% water, 20% organic matter and 70% inorganic salts. Cementum surrounds the dentine, and both contain collagen. Enamel has only 3% water and 1% organic matter – the rest is HA. Keratin is the protein in the enamel. The mineral phase (HA) is 97% by weight in enamel and 69% by weight in dentine (volume-wise it is 92% and 48% respectively). Weight-wise, the organic phase is 1% in enamel and 20% (18% collagen, and rest citric acid, mucopolysaccharides, insoluble protein and lipids) in dentine. Water content is 2% and 11% respectively, by weight, in enamel and dentine.

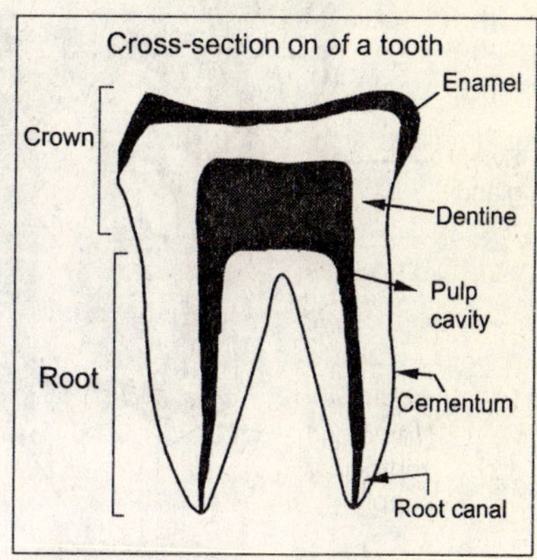

Fig.11.2 Structure of tooth

Of the connective tissues, bone and cartilage give protection and support, blood is essential for life, while others produce loose or dense fibrous tissue and adipose (fat) tissue.

11.7 Skin

The skin or cutis (Fig.11.3) is divided into the superficial *epidermis* and deeper *dermis* (*cutis vera* or true skin). Stratified epithelia of various thickness forms the epidermis (thickest in palms and soles), which has four layers – most superficially stratum corneum (horny layer), stratum lucidum (clear layer), stratum granulosum (granular layer) and deepest stratum germinativum (basal layer) – often the deepest polygonal tall columnar or long cuboidal cells form a fifth stratum cylindricum, resting on the basement membrane over the dermis. The dermis varies from 0.2 to 4 mm in thickness and contains collagen, fibroblasts, histiocytes and pigment cells, sweat glands and sebaceous glands. Appendages of the skin include hairs and nails.

11.8 Further Reading

[1] Ganong, W. F. *Review of Medical Physiology* 20th ed. Prentice Hall Intl (Inc.). 2001.

[2] Guyton, A. C. and Hall, J. E. *Textbook of Medical Physiology* 10th ed. W B Saunders. 2001.

[3] Elderton, R. J., ed. *Clinical Dentistry in Health and Disease.* Dentition and Dental care, vol. 3. Oxford: Heinemann. 1990.

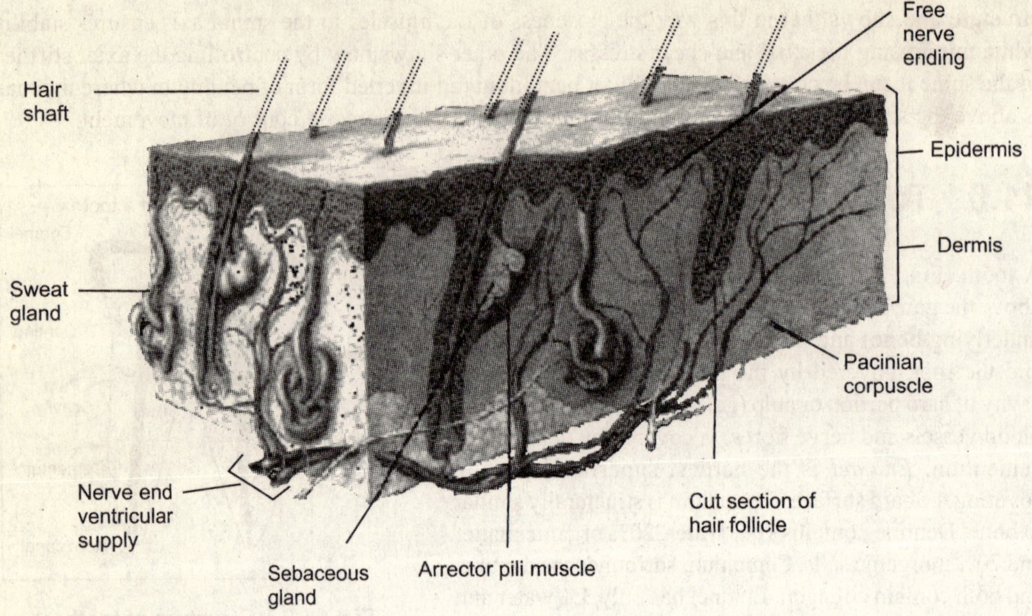

Fig.11.3 Structure of skin

[4] Skakle, J. M. S. and Aspden, R. M. Neutron diffraction studies of collagen in human cancellous bone. *Journal of Applied Crystallography* 35: 506-508. 2002.

[5] Aspden, R. M. et al. *In vitro* behaviour of human osteoblasts on dentine and bone. *Cell Biology International* 26: 337-346. 2002.

[6] Aspden, R. M., Scheven, B. A. A. and Hutchison, J. D. Osteoarthritis is a systemic disorder involving stromal cell differentiation and lipid metabolism. *Lancet*, 357: 1118-1120. 2001.

[7] *www.abdn.ac.uk/medicine_therapeutics/staff/ralston*

[8] *http://animaldiversity.ummz.umich.edu*

[9] *http://bioweb.uwlax.edu/zoolab*

[10] *http://www.npaci.edu/envision/v15.3/keaveny.html*

11.9 Questions

1. Descriptive Questions

1. What do you understand to be connective tissue ?
2. How do bones and teeth differ from other connective tissues ?

2. Write Short Notes on

1. Collagen
2. Hydroxy apatite (HA)
3. Enamel

4. Elastic cartilage
5. Fibrous tissue

3. Multiple Choice Questions

Choose the single best response.

1. Man can tolerate without serious damage to his muscles, a positive G force up to
   (A) 100G
   (B) 10G
   (C) 7G
   (D) 4G

2. Which is not a connective tissue?
   (A) myoglobin
   (B) blood
   (C) fat
   (D) teeth
3. The white fibrocartilage is not
   (A) tough
   (B) incompressible
   (C) resistant to stretching
   (D) shock absorber
4. Modelling of tissues as fibre composite materials shows how tapered fibers, are advantageous for
   (A) maximising stress transfer
   (B) minimising the metabolic cost of the materials
   (C) both
   (D) none
5. Mechanical loading of cartilage and bone tissue, associated with daily activities, results in
   (A) movement of fluid contained within the tissue
   (B) enhanced transport via convection.
   (C) both
   (D) none

Answers

1. D; 2. A; 3. B; 4. C; 5. C

12

Properties, Compatibility Characteristics and Performance of Implants

12.1 Introduction

When rehabilitating an affected individual to normal life, the prosthetist-orthotist is responsible for evaluation, design and fabrication of *orthoses* (braces or orthopedic appliance or apparatus used to support, align, prevent or correct deformities or to improve the function of the movable parts of the body, from the Greek word meaning 'making straight',) or *prostheses* (an artificial substitute for a missing body part, viz., arm or leg or eye or tooth or heart valve or cochlea, used for functional or cosmetic reasons or both, from the Greek word meaning 'a putting to'). His job is to see that the device fits and functions properly and that the patient adjusts to its presence. However for cardiac or neural prostheses, the concerned medical specialists usually decide.

The properties and compatibility characteristics of biomaterials are described in Chapters 8 and 13.

12.2 Orthoses

Orthoses are usually described by the joints they encompass and the mechanical motions they purvey. Although they are useful in protecting and immobilising the joint, their common drawbacks include the difficulty in obtaining patient compliance (inversely proportional to the degree of immobilisation), the difficulty of donning and doffing the device, skin breakdown, nerve compression, muscle atrophy, restriction of the movements of the affected area like chest or throat or abdomen. Non-physiological negative effects of some orthotics, especially temporary spinal orthoses, are usually concerned with the cosmetic and psychological aspects, including anxiety about altered appearances, emotional dependence on the device and embarrassment.

Orthoses require external compression on the skin molded to the body contours. To achieve fixation, skin and body habitus must be considered. Overweight patients may be difficult to fit. Patients with insensate (anesthetic) or fragile skin, secondary to disease, are prone to skin breakdown. These are important factors to be considered in orthotic design and decision of the fabrication material to be used.

12.2.1 Orthotic Materials

A multitude of materials have been used for fabricating orthoses. Traditional ones include metal, rubber, leather and canvas. Recent additions include plastics and synthetic fabrics. The important factors for considering their use would be strength, durability, flexibility and weight. The design should be simple, comfortable and as aesthetic as possible. It must distribute forces over a sufficiently large surface area. Parts in contact with the body must be properly contoured and padded.

Conventional orthoses use metals to purvey strength and durability, with straps and padding made of leather. Commonly used metals include steel, aluminum and their various strong, corrosion-resistant alloys. Despite the heavy weight and unattractive appearance, metal orthotic devices are adjustable, and therefore, can accommodate for growth and altering requirements.

Plastic orthoses are lighter and fit more closely as they can be molded directly to the body or a plaster replica of the body part. Close fits of these purvey wider distribution of corrective forces compared to a metal prosthesis. Adding foam liners inside will add to the comfort. Usually they are adequately strong and durable, albeit not adjustable in length. Some plastics, however, allow reshaping when heated to accommodate or purvey relief at pressure points. Normally they are fitted with metal joints or spring-loaded plastic bars since plastic joints do not last long.

Plastics used in orthotics belong to two main groups: thermosetting and thermoplastic materials. Thermosetting (e.g., formaldehyde, epoxy and polyester resins) materials are commonly used as laminates. They need heat for initial hardening, but they do not soften with subsequent heating. Thermoplastics, on the other hand, soften whenever they are heated, making the material mouldable. Lowering the temperature hardens the material once more. Low temperature thermoplastics, viz., orthoplast and plastazote, become malleable just above body temperature. This permits quick fabrication and moulding directly onto the body. However, since these materials are not very strong or durable, they are not recommended for long-term use. High temperature thermoplastics (e.g., polyethylene, polypropylene, copolymers, ortholene and vinyl polymers) require heating up to 150°F or more to make them workable. Fabrication of orthoses with any of these materials requires an exact plaster replica of the body part. The heated plastic is then applied to the replica for proper moulding. These materials are strong, durable and elastic, returning to their original shape after flexible deformation. These purvey support and may also give a spring-action assist force. Most long-term plastic orthoses are made of high temperature thermoplastics.

Indian researchers have built a Floor Reaction Orthosis, a 300-gram calipers (made up of missile material) and test-driven them in orthopedic centres. These orthoses have been a great help to poliomyelitic children - instead of dragging around a three-kg load (of metallic calipers) on their legs, they can now move around very easily!

12.2.2 Spinal orthotics

The spine may be divided into six parts from the cephalic end (head) – the upper cervical ($C_{1,2}$), the mid-cervical (C_3–D_1), the thoracic or dorsal (D_{2-10}), the thoracolumbar (D_{11}–L_1), the lumbar (L_{2-4}) and the most caudally lumbosacral (L_5–S_1) part. The two extreme ends are the most difficult to immobilise. Spinal orthotics may cause skin irritation, osteopenia (rarefaction of bone), muscle atrophy and psychological problems, including emotional dependence. The orthoses may be classified as cervical (CO), head cervical (HCO), cervicothoracic (CTO), cervicothoracolumbosacral (CTLSO), thoracolumbosacral (TLSO) and lumbosacral (LSO).

12.2.3 Other orthotics

The common lower limb orthoses are the AFO (ankle-foot orthosis), the KAFO (knee-ankle-foot orthosis) and the KO (knee orthosis). Spinal orthoses have already been mentioned in Section 12.2.2. Another common and important orthosis (of the upper limb) is the thumb spica (for thumb carpometacarpal/CMC stabilisation), which is a static, functional splint that immobilises the wrist and the CMC joint of the thumb, while allowing interphalangeal (IP) joint motion.

12.3 Prostheses

Prostheses can be functional, cosmetic or both. Prosthetic fitting (of a limb) can be done once the residual (amputated) limb is ready for casting. During fabrication, a temporary, provisional or preparatory prosthesis is first made. After the shaping and shrinking process of the residual limb is complete, a definitive or permanent prosthesis is made. Nowadays, for the preparatory prosthesis, thermoplastics are used more than conventional or synthetic plaster bandages. Most thermoplastic sockets are custom made and better fitting. As the plastic needs an oven heating up to 300°F or more, for moulding, the socket must be formed over a positive mould of the residual limb. After modifying the positive mould, the provisional socket must be vacuum formed, often over a foam liner. Nowadays, pre-formed sockets made of a carbon fibre and fibreglass-reinforced water-activated polymer, allow a light-weight durable socket to be moulded directly on to the amputee's residual limb, within hours.

12.3.1 Limb Prosthetic Materials and Methods

Now CAD-CAM techniques are used either as an adjunct or an alternative to traditional socket fabrication. These can be used to get a digitised image of the residual limb or a cast of the residual limb (negative mould) directly to purvey a positive computer model of the residual limb. This image is then modified – removing plaster from pressure-tolerant areas and adding plaster to pressure-sensitive areas of the residual limb. After which, a computer-controlled milling machine will create a positive model for creating the check socket. Without a clear check socket, the final prosthetic socket may be fabricated directly over the modified plaster positive mould and fitted on the patient – making additional modifications directly to the final socket. A clear thermoplastic sheet (e.g., lexan, duroplex, surlyn) or cone (e.g., orthoglass) is vacuum formed over the modified positive and the fit is then checked on the patient.

As the check socket is clear, areas of increased pressure or of non-contact between the socket and the residual limb will be seen on direct observation. The fit can be modified quickly to achieve desired optimum fit of the socket and a new positive mould can be made for fabrication of the definitive socket. A thermoplastic check socket can be used with a suspension system and other components attached to the temporary prosthesis, which in turn, can further evaluate the fit and the resulting function, or adjust placement of other components like electrodes in myoelectric prostheses.

Most definitive prosthetic sockets and external supporting frames are still fabricated from thermosetting resin – polyester or acrylic. Resins are drawn under vacuum into a fabric stockinette (commonly dacron or nylon) reinforcing the plastic. Other reinforcing fabrics include graphite fibres and Kevlar, which purvey very light but strong laminations.

The construction can be exoskeletal (traditional – rigid plastic lamination over a filler of wood or foam – shaped cosmetically), or endoskeletal (internal modular components and tubing purveys strength for weight-bearing capacity and cosmetic shaped foam covers slipped over them). The endoskeletal ones can become even lighter if thermoplastic sockets and titanium or graphite modular components are used. They may be passive or powered (external electric motors or switch controlled or myoelectric).

In the case of lower extremities, TES (total elastic suspension) or belt suspension prosthetics are also used. Another example of common lower limb prostheses is SACH or solid ankle cushion heel prosthesis.

Jaipur foot

The Jaipur foot is the key element of the Jaipur limb, an artificial leg developed over the past 20 years by Dr P K Sethi, an orthopedic surgeon in Jaipur. As head of orthopedics at a hospital there, he noticed that more and more amputees were disinclined to use the (Euro-American) artificial limbs they had been fitted with, but were, instead, going back to using crutches. So, Dr Sethi developed an artificial limb that differed in a number of ways from the Euro-American model. It had a foot, which looked like a foot instead of a piece of machinery to be concealed by a shoe. It was made waterproof by encasing it in easily available rubber. He then replaced the single wooden heel of a Euro-American artificial foot with two wooden blocks linked by a large rubber block. This made the foot flexible enough for squatting. Finally the material for the front block which was originally wood was replaced with hard rubber, giving the front of the foot more mobility and making it easier to climb hills. The result was not only a design more suited to less developed countries, it was actually a better design, as was shown when the foot was tested by Strathclyde University: the amputee who tested it refused to hand it back because he found it so much better for climbing the Scottish mountains than the Euro-American model. Unsurprisingly, several of the foot's features are being incorporated in products in Europe and USA. Although, it is heavier than the Euro-American equivalent, and not so easy to fit into a shoe, it is simple, cheap and with its aluminum limb (another readily available material), effective.

There is a second and rather similar example of how the Jaipur team developed a better technical solution to a medical problem than the West's: the calipers fitted to the legs of polio sufferers. Since the disease has been effectively eradicated in the West, the design of the caliper (with its two metal side bars) to support the damaged knee has hardly changed in a century. To achieve better mobility, the hospital developed a support made out of polymers and carbon fibre, moulded to the leg and attached by Velcro fasteners. It was four times lighter than the metal alternative and allowed movement of the knee while still supporting it. Carbon fibre has to be imported, but many elements of the product can be made by local craftsmen: indeed, on a pilot project they improved on the design.

12.3.2 Powered Upper Extremity Prostheses

Powered prostheses may be body powered or externally powered. Body powered upper extremity prostheses consist of the parts given in Table 12.1.

Table 12.1 Parts of a body powered prosthesis

	Name of part	Types
1.	Terminal device(s)	Hooks, functional hands, cosmetic hands (covered with skin colored gloves), special (*e.g.*, bowling ball or golf club holder)
2.	Interposing joints	As required by the level of amputation
3.	Socket	Conventional, shoulder saddle with chest strap, suction or self-suspending
4.	Suspension	Figure 8 or figure 9 harness
5.	Control system	e.g., Bowden double control cable

Of the terminal devices, functional hands are heavier than hooks. The cosmetic hands are passive. In this group, voluntary opening of the hand is by springs or rubber bands that purvey prehensile forces. The voluntary closing devices allow the user to provide and control a variable prehensile force transmitted through the control cable to the terminal device. These also purvey greater proprioceptive input by means of force exerted on the control cable.

An example of an interposing joint for the wrist unit (of the hand) will be the receptacle connecting the terminal device to the prosthesis and pre-positioning it for its function (like supination or flexion). The rotation may be achieved by friction or a locking mechanism.

The conventional upper extremity socket is usually suspended by a figure of 8 harness, to which the control cables are attached. More loads can be accommodated in the case of a shoulder saddle with chest strap. Suction or self-suspending sockets (e.g., Muenster) purveys partial or complete suspension with a figure of 9 harness.

Switch-controlled prostheses, externally powered devices, have small micro-switches incorporated either inside or outside the prosthesis socket and are operated on contact by the amputee (patient). Pull switch within the conventional harness and cable control systems are also available for controlling prosthetic function.

Another variety of externally powered prostheses is the myoelectric type. It is the most popular one, with a self-suspending socket (without harness). It has set(s) of electrodes embedded in the socket. These are in contact with the skin and can detect muscle action potentials from a voluntarily contracting muscle in the residual limb. This signal is amplified and rectified. The final signal turns on an electric motor to purvey function (e.g., terminal device operation of wrist rotation or elbow flexion). The various systems described below can process myoelectric signals.

(i) Two sets of electrodes and amplifiers to control motion in one direction, and one for the opposite direction (e.g., terminal device opening and closing).

(ii) In the single site system, one set of electrodes is used for both motions. Here the strength of voluntary contraction dictates the final motion. To cite an example, weak contraction closes and strong contraction opens the terminal device. This type of control system allows only one good control site and may allow 2 control sites to power 2 different functions, as in the case of terminal device opening and wrist rotation.

(iii) In some systems, the strength of contraction controls the speed of a function. For instance, the Liberty Mutual or Boston elbow is a proportional electrical elbow that extends or flexes faster in direct proportion to greater amplitude of contraction within the control muscles.

(iv) The most sophisticated Utah elbow uses microprocessor technology and 2 sets of electrodes to purvey both elbow function and terminal device operation. It permits free-swing phase of elbow joint for purveying normal cosmetic arm swing during walking.

However, the disadvantages include cost, reliability and weight.

12.3.3 Artificial Joints

Artificial knee joints are implanted in patients with a diseased joint to alleviate pain and restore function. After about 10 years of use, these artificial joints often need to be replaced because of wear and fatigue-induced delamination of the polymeric component. Now, engineers are developing improved materials to extend the lifetime of orthopedic implants such as knees and hips. The Indian Prime Minister, during the turn of the millennium, is a widely publicised recipient of an artificial knee.

The Jaipur limb, originally developed in Jaipur, India, is perhaps the most successful and widespread, culturally appropriate and economically sustainable prosthesis. It is basically a soft multiaxial SACH foot, meaning that it is flexible along multiple axes and has a solid ankle, cushioned heel. It is known for its durability, waterproof qualities, and dorsiflexion (the ability to point the toes up in the air, a movement essential to squatting).

The Jaipur limb is an economically feasible opportunity to many amputees due to funding from the Indian government as well as many local organisations. It is culturally appropriate because it allows users to engage in everyday physical activities in rural India, such as squatting (an important body posture in India for resting, sitting, and using toilets), walking barefoot or with thongs over rough terrain, and submersing the prosthesis in water (for washing feet or walking in farm fields during the rainy season). Aside from appropriateness to local culture, another interesting aspect of the Jaipur limb is that it is also locally manufactured.

But behind the scenes of making feet is the equally interesting process of making the moulds used in making the feet. The foot, itself, is probably the better known component of the Jaipur limb, but the socket is just as important. Originally made of aluminum, but now also of PVC irrigation pipes and aluminum in tandem with polypropylene in various versions, the socket is a crucial element of the prosthesis. Most versions are open-ended plug sockets which bear the amputee's weight on a popliteal shelf and patellar tendon bar (the upper calf muscle and just below the kneecap). The most common suspension system, or method of affixing the prosthesis, is a leather cuff above the knee.

On assessing the present scenario towards improving design, functional requirement and aesthetics of the artificial limbs in India, the Advanced Composite Mission of TIFAC launched a project on developing 'composite artificial limbs for the physically handicapped' [*http://www.tifac.org.in/do/acm/case/limb.htm*] in collaboration with M/s. Mohana Orthotics and Prosthetic Centre, Chennai (Madras). The Department of Aeronautical Engineering, Madras Institute of Technology (MIT), Chennai has been extending technology support in terms of design, prototype development and complete testing of composite limbs. The project activities involve mechanical design and its validation, prototype development and productionisation by suitable fabrication techniques and equipment. The below-the-knee endoskeleton artificial limbs developed under the project are lighter in weight and have a better appearance than ever before with improved gait for the patients. This below-the-knee endoskeleton limb consists of five parts: a FRP tubular structure fabricated by filament winding of glass fibre in epoxy matrix, top and bottom connectors made by injection moulding of glass filled nylon, a polyurethane foot with a composite keel embedded in it and a polypropylene socket to accommodate the amputee stump. The socket made of polypropylene is patient- specific and does not create any problems like pressure sores even for diabetic patients. The FRP tube connects the socket to the foot. The connectors between the socket and tube and tube and the foot can be adjusted for angular alignment of the limb. A very innovative design approach has been adopted for designing FRP keel for providing improved strength and flexibility in the foot piece. All the components of the limb have been designed on the basis of theoretical analysis using CAD software (CSA/NASTRAN) for typical compression loads at different angles, momentary impact etc. All the five parts and the socket are adjustable to meet individual requirements and to take care of static and dynamic alignment patterns. The evaluation of individual components and also of the entire endoskeleton assembly for compressive and bending strength has been carried out. A simulated endurance test has been conducted for a 5-year service life of the artificial limb considering average stance duration of 0.5 seconds for normal gait and three hours walking time per day thus testing the limb for 39.42 million cycles to simulate the life of 5 years. A 3-D modelling of the endoskeleton with all the embedded components considering actual properties of various materials was also carried out. Such a type of below the knee artificial limbs would cost only Rs.5000–6500/- only as against Rs.40,000/- for the imported ones.

12.3.4 Cardiac Prostheses

In this section, we shall broadly discuss two groups of cardiac prostheses viz., heart valves and cardiac pacemakers.

Heart valves

Diseased heart valves may be replaced by the patient's own valve (autograft, e.g., pulmonary for aortic), bioprostheses (heterografts – porcine or bovine; and homografts – from other human donors) or mechanical valves (St. Jude®, Starr-Edwards®, Medtronic Hall®, Biopro®, Chitra®). Autograft has the least complications but requires a very high-order skilled surgery. Bioprostheses have low thromboembolic potential and, in many cases, anticoagulants are not needed. The major disadvantage is its limited durability (after 10 years, the failure rate rises quite sharply). Another disadvantage is the modest outflow obstruction and residual pressure gradient in small valve surgery. Homografts are not easily available but are relatively resistant to bacterial endocarditis. In the rare case of isograft (from an identical twin) there is no chance of graft rejection or reaction by the host, because of the identical genetic composition of the host and donor tissues. All mechanical valves require anticoagulation, but they are more durable than bioprostheses.

Among mechanical cardiac valve prostheses, the St. Jude bileaflet valve has low rates of thromboemoblism and long-term durability. St. Jude has improved this technique and has developed a wide range of devices based on their bileaflet technology. The mechanical components are made of pyrolytic carbon. This mechanical valve is positioned in the aortic area. It is sutured just above the fibrous ring, at the base of where the natural aortic valve was located. This eliminates interference with the flow of blood into the coronary arteries. The BioImplant tissue valve consists of a porcine (pig) aortic valve that has been fixed in glutaraldehyde and then sewn into a stent (rigid frame) surrounded by a sewing ring. Researchers have suggested that biological cardiac prostheses may last longer and work better without a stent, allowing them to deform naturally. An example is given below.

The Toronto SPV valve is a more recent addition to St. Jude's tissue product line. It is a porcine aortic valve that has been chemically treated to stabilise the tissue, improving the durability and reducing the immune response in the patient. Unlike many earlier biological valves, the Toronto SPV is stentless, containing no external rigid frame. The intention is to allow the valve to deform properly during opening and closing. St. Jude has also introduced the SJM X-Cell bioprosthesis, which is a porcine valve that has had the pig's cells biochemically removed. The theory is that by eliminating the pig cells, the rate of calcification within the valve will be reduced. Insufficient evidence exists to determine if this approach improves long-term durability.

Table 12.2 Technical specifications of Biopro

External Diameter (mm)	Internal diameter (mm)	Height (mm)
19	14	10
21	16	11
23	18	12
25	20	13
27	22	14
29	24	15
31	26	16
33	28	17
35	30	—

The cardiac valvular prostheses using bovine pericardium, Biopro PB (comprises a delrin support having three shafts coated with bovine pericardium over which the biological structure is mounted. The pericardium is treated with 5% glutaraldehyde solution, which permits preservation among collagen bundles and tissue elasticity. A special stainless steel wire inserted externally in the support ring reinforces the ring's base consistency and allows identification of the patient's bioprostheses by simple radiological (X-ray) examination. The technical specifications of Biopro are given in Table 12.2. Biopro PB may be used for replacement of mitral, aortic, tricuspid and pulmonary valves by directing the cardiac blood flow. It may also be utilised in valvulated tubes from bovine pericardium for extracardiac diversions.

Chitra Heart Valve

India's first indigenous development – the Chitra Heart Valve – a tilting disc, mechanical heart valve prosthesis is a culmination of more than 15 years of painstaking research and development. The technology was conceived and crafted by the Sree Chitra Tirunal Institute of Medical Science and Technology, Thiruvananthapuram and funded by the National Research Development Corporation, Government of India, New Delhi.

The valve holds a lot of promise for persons suffering from rheumatic heart diseases (blocked or leaking aortal and/or mitral passages in the heart). It has been tailored in four different sizes for aortal and mitral passages in the heart. The heart valve has a cage made of a special implant material with good tissue acceptability and high-wear resistance. The tilting disc is fabricated from ultra high molecular weight polyethylene (UHMW/PE) and the sewing ring is knitted from special cloth.

After successful multi-centric clinical trials at six different hospitals in India, TTK Pharma, has commercially launched the 'Chitra heart valve', which is available at an affordable price. Further, the technology has been patented in the US, India and patent applications are pending in the European Patent Office (6 countries) and Japan.

Artificial heart valves implanted in patients in India (many of them are children who suffer from rheumatic fever) till a few years back were imported at a price which range between three to six times that of the Chitra valve. The Chitra heart valve is priced at around US$ 350. In the Indian context, the durability of the valve is of paramount importance because the average age of RHD patients is between 20 and 30 years, whereas in the developed countries patients are mostly above sixty. Expected to serve for at least 20/25 years, the Chitra valve has been tailored for local patients who have smaller hearts than their foreign counterparts. It is available in sizes between 21 and 27 mm (sewing ring diameter or maximum space available for the valve), and meets more than 95 per cent of the implant requirements in the country. Other rarer sizes of 19 and 29 mm will also be available shortly (Table 12.2). The Chitra valve has a circular cage which supports a tilting disc with an opening angle of 70°. Unlike the standard disc in the Bjork-Shiley valve, a popular international tilting disc model, the Chitra valve has a 'taper' (smooth peak) and a 'well' (gentle slope) located on the outflow side of the non-seating disc, which allow improved blood flow through small orifice areas. Since heart valves are to function continuously and the disc which is tiltably mounted has to flutter more than 115,000 times per day in a normal person having 72 to 80 beats per minute, the artificial disc of a tilting disc valve has to open and close 40 million times a year at an average, causing tremendous strain on the material. So the material used has to be extremely durable in terms of wear resistance and fatigue strength. A valve may have to operate inside the human body up to 25 years or more which necessitates the material to have excellent chemical and structural stability. The biocompatibility (both tissue compatibility and blood compatibility) of the material is also a major requirement to ensure acceptability of the implant by the body. Finally, the material should be processable to achieve the required surface finish and dimensional stability.

A modified form of ultra high molecular weight polyethylene (UHMW/PE) is best suited for the purposes of heart valve applications since it exhibits excellent tissue and blood compatibility and is stable (both chemically and structurally) for long duration of working in physiological environments. It has extremely good wear resistance, fatigue strength and toughness. The improved heart valve disc using modified UHMW/PE will ensure that the moulded heart valve disc is given a high surface finish so that the tendency of clot formation by the flowing blood on the disc surface is substantially eliminated. It is required to have a surface finish of the order of 0.1 μm or better. UHMW/PE also has all the other advantageous properties such as durability, longevity, high stress resistance, wear resistance and blood compatibility. The heart valve disc is prepared involving a solid state compaction technique for generating the required surface finish along with a cryo/machining technique for flash removal. The details are beyond the scope of this book

Design features

The heart valve assembly (Chitra) comprises

1. A sewing ring component
2. A metallic valve housing component, and
3. A disc occluder.

The disc occluder is held tiltably on supports within the enclosed space of the valve housing component. It has a substantially flat surface on one side and a substantially tapered surface on the opposite side. On the tapered side, it possesses a centrally depressed portion, for engaging it on the valve housing. The disc occluder is optionally purveyed with a marginally depressed portion on the substantially flat surface, the depressed portion being not as prominent as the depressed portion of the other side of the disc. The circular valve housing has an externally grooved portion for the sewing ring component, and the disc occluder is mounted within its inner circumference tiltably on supports within the enclosed space. The sewing ring component is fabricated from polyester, the valve housing is fabricated from a special alloy, while the valve disc is fabricated from modified UHMW/PE. The fabrication of the disc and the modification of the UHMW/PE take place simultaneously.

The reader is strongly encouraged to do a Google Search under the "Images" category [http://images.google.com] to see the figures of all these implants crafted by the various manufacturers.

Cardiac pacemakers

As discussed in Chapter 2.2, the heart has its intrinsic pacemaker activity originating from the sino-atrial (SA) node. When some disease upsets the origin of this rhythm, (artificial) cardiac pacemakers become necessary for sustaining life.

Cardiac pacemakers are temporary (transcutaneous or transthoracic) and permanent. The latter has a complex pacing system and is usually denoted by 3 letters – the first for the paced chamber (A/atrium, V/ ventricle, D/double or both), the second letter for the chamber sensing the electrical signals (A, V, D or O/ neither) and the third letter for the mode of response (I/inhibitory, T/triggered, O/not applicable). Sensor-based pacing systems (rate modulation according to the metabolic need) add an R as the fourth letter. The other letters which may be placed in the fourth place are O (none), P (simple programmable), M (multi-programmable), and C (communicating). There may occasionally be a fifth letter denoted by O (none), P (packing), S (shock) and D (dual or P+S). The three-lettered types are exclusively used in bradyarrhythmias i.e., when the heart rate is persistently below 60 bpm (beats per minute). The five-letter type is exclusively used for tachyarrhythmias i.e., when the heart rate is persistently above 100 bpm.

Modes of pacing may be asynchronous (VOO, AOO or DOO) or single chamber demand pacing (VVI, AAI, VVT, AAT), P-synchronous (VDD) or universal (DDD). Unipolar pacing systems have the cathode (stimulating electrode) in the heart and the anode at the generator. The distance between the two cause large pacing artifacts whose axis in the frontal plane points towards the anode. In a bipolar pacing system, both the electrodes are in the heart (1–2 cm apart) in either or both atria and ventricle. The distal (tip) or proximal (ring) electrode of the lead can serve as the cathode. As the inter-electrode distance is small, pacing artifacts are also small and their axis in the frontal plane reflects direction of current flow.

Cardiac (artificial) pacemakers are commonly made with CMOS (complementary metallic oxide semiconductors).

12.3.5 Dental Implants

Other than the limbs and the heart, one area of our body which is frequently (probably most commonly) subjected to implants is our teeth.

Dental implants are dealt with by orthodontics, a branch of dentistry. The implants may be classified as endosseous (metal or ceramic in the mandible/maxilla), subperiosteal and intramucosal. The materials for its manufacture can belong to any of the following groups, viz., polymers, metals, ceramics or composites. Among the metals, Titanium (Ti) – being the most corrosion resistant and flexible is very much preferred. Tantalum is also popular. An alloy of cobalt, calcium and molybdenum (vitallium) is also used now. As toxicity of aluminum is not yet widely publicised, alumina (Al_2O_3) and porous calcium aluminum silicate ($CaAlSiO_3$) are still used. Other materials include vitreous carbon, calcium phosphate or $Ca_3(PO_4)_2$ and hydroxyapatite (HA).

Synthetic teeth made up of polymer/ceramic composites (with microfine filler particles of silica) are better than acrylic teeth for wear resistance, and are superior to porcelain teeth in bonding to denture base materials.

Orthodontic appliances can be either removable (polymeric framework) or fixed. The latter may be either bands cemented around a clinical crown or brackets directly bonded to acid-etched enamel. Orthodontic base polymers are of two forms: powder (PMMA, PEMA and peroxides) and liquid (MMA with 0–6% ethylene glycol dimethacrylate and amines). Commonly corrosion resistant alloys like stainless steel, Co-Cr, Ni-Cr, Ni-Ti, b-Ti and Au are used. However, nickel is allergenic in many people and gold is too soft. Titanium alloy is stable at room temperature (79% Ti, 11% Mo, 6% Zr and 4% Sn by mass). HA is also being used with Ti for natural bonding.

Contact lens

PMMA (poly methyl methacrylate) is also used for 'hard' (conventional) contact lens of eye because of its optical properties (RI: 1.496), chemical inertness, lack of clinical toxicity, weight and ability to be machined. Newer polymer variants have greater wetability and differing gas transport properties but PMMA remains the standard material for contact lens. The contact lens rests on the tear layer bathing the cornea. Dissolved oxygen in tears is necessary for aerobic metabolism in corneal epithelium. If tear interchange is inadequate, corneal glycogen stores will be depleted.

On the other hand, hydrophilic or 'soft' contact lens is made up of HEMA (hydroxy ethyl MA) that is also a polymerised plastic. Other cross-linked and copolymeric materials include ethylene glycol dimethacrylate (EDMA) and PVP (poly vinyl pyrrolidone), which are normally rigid and brittle, but with 30–85% water, become pliable.

12.3.6 Neuroprostheses

Neuroprostheses include cochlear and vision implants.

Before we discuss neuroprostheses, we must start with a note on hearing aids – the more common prosthetic appliance for the hearing impaired. It can be of various types, like, body hearing aids, behind the ear (BTE) hearing aids, in-the-ear (ITE) hearing aids, in-the-canal (ITC) and completely in-the-canal (CIC) hearing aids, programmable hearing aids and contralateral routing of signals (CROS) hearing aids. Of these, the last one is meant for the unilateral deaf (whose other ear is normal or working). Essentially all these act as mere amplifiers, and cannot be used when the neurons of the inner ear (cochlea) are greatly damaged.

The cochlear implant (artificial ear) comprises the following components.

i. Small microphone.
ii. Long cable.
iii. Speech processor.
iv. Cable.
v. Transmitting coil.
vi. Receiver/Stimulator (R/S).
vii. Array of electrodes.

(i) (in the above list) picks up the environmental sounds, sending it through (ii) to (iii). (iii) is nothing but a miniature computer, which filters, analyses and digitises sound into coded signals. These coded signals are sent via (iv) from (iii) to (v). (v) sends the signals across the skin to the implanted R/S (vi) through FM radiosignals. (vi) delivers appropriate electrical stimuli to proper electrodes on the array (vii). Finally, (vii) stimulates the remaining nerve fibers in the cochlea. This is an example of a multi-channel FES of the nervous system.

As far as artificial eye is concerned, it is probably much further away. An array of 100–1000 electrodes in the visual cortex (occipital lobe, area 17), can produce visual perceptions as in an electronic scoreboard. A miniature video camera, worn by a blind person, with proper processing, can give street names and outlines of environmental structures fairly well. Nevertheless, biomaterial compatibility or biocompatibility, for long-term implantation in the brain, is a major obstacle.

In the area of neuroprostheses also (Bio-)MEMS (Section 1.4) is being used nowadays as a sensor.

12.4 Further Reading

[1] Furman, S., Hayes, D. L. and Holmes, D. R. Jr. *A Practice of Cardiac Pacing*. Futura. 1993.

[2] McCullough, N. C. III. *Atlas of Orthotics.* St. Louis: C. V. Mosby. 1975.

[3] Esquenazi, A. ed. *Prosthetics – State of the Art Reviews*. PA: P M & R. Hanley & Belfus. 1994.

[4] Kominsky, S. J. ed. *Medical and Surgical Management of the Diabetic Foot.* St. Louis: C. V. Mosby. 1994.

[5] Braddom, R. L. ed. *Physical Medicine and Rehabilitation.* PA: W B Saunders. 1996.

[6] Sarmiento, A. and Sinclair, W. F. *Atlas of Orthotics: Biomechanical Principles and Applications.* St. Louis: C. V. Mosby. 1975.

[7] Bowker, J. H. and Michel, J. W. *Atlas of Limb Prosthetics: Surgical, Prosthetic and Rehabilitation Principles.* American Academy of Orthopedics Surgeons. St. Louis: C. V. Mosby. 1992.

[8] Perier, P., Besou, J., Swanson, J. et al. Comparative evaluation of aortic valve replacement with Starr, Bjork and porcine valve prostheses. *Circulation* 72: 140. 1985.

[9] Graber, T. M. and Swain, B. F. eds. *Current Orthodontic Concepts and Techniques*, 2nd ed. PA: W B Saunders. 1975.

[10] Williams, D. F. ed. *Concise Encyclopedia of Medical and Dental Materials.* Oxford: Pergamon Press. 1990.

[11] Craig, R. J., O'Brien, W. J. and Powers, J. M. *Dental Materials: Properties and Manipulation.* 6th ed. Harcourt Brace & Co. Asia Pte. Ltd. 1998.

[12] O'Malley, T. J. A review of the functional electrical stimulation equipment market. *Assist. Technol.* 4: 40–45. 1992.

[13] RESNA. proc. RESNA'96 Conference Supplement: Research Symposium: Recent developments in neuroprosthetics: progress towards functional artificial vision. Washington DC. 1996.

12.5 Questions

1. Descriptive Questions

1. How do prosthetics and orthotics help in rehabilitation ?
2. What are cardiac prostheses ?

2. Write Short Notes on

1. Prosthetist-orthotist
2. Spinal orthoses
3. Limb prostheses
4. Dental implant materials
5. Neural prostheses

3. Multiple Choice Questions

Choose the single best response.

1. Orthotic materials may be made up of

(A) metal
(B) leather
(C) plastic
(D) either

2. Plastic material for orthotics may be all except
(A) polyester resins
(B) copolymers
(C) vinyl polymers
(D) none

3. Arrange in order of preference the heart valves causing least complications (I-autograft, II heterograft, III-homograft, IV-artificial)
(A) I-II-III-IV
(B) IV-III-II-I
(C) I-III-IV-II
(D) I-IV-III-II

4. Universal pacing of a cardiac pacemaker is denoted by
(A) DDD
(B) DOO
(C) AOO
(D) VOO

5. The most preferred metallic dental implant material is
(A) Tantalum
(B) Titanium
(C) Vitallium
(D) Aluminum

Answers

1. D; 2. D; 3. C; 4. A; 5. B

13

Structural Properties of Synthetic Biomaterials and their use as Biomaterial Substitutes

13.1 Introduction

As mentioned in Chapter 11, a biomaterial is any substance (other than a drug), synthetic or natural, that can be used as a system or part of a system that treats, augments, or replaces any tissue, organ or function of the body.

Biomechanics is the science of forces acting internally or externally on the living organism or the cell. It may also be defined as the application of mechanical laws to living structures, as to a locomotor system (*cf.* Chapter 14). The relationship between the behaviour of biological structures and the physical influences of a biomaterial or prosthesis may also be studied (*cf.* Chapter 15).

The ideal properties for a (synthetic) biomaterial include being: non-toxic, non-irritant, non-inflammatory, non-allergenic, non-mutagenic or carcinogenic, chemically (biologically) inert, corrosion resistant, mechanically-resistant to fatigue failure, easily sterilisable and economical.

13.2 (Bio) mechanics of Biomaterials

The biomechanical properties of (metallic) wires include great flexibility (lower and more constant force over a greater distance), which is a resultant of both their mechanical properties and their geometry. They also have high resilience (ability to store and release elastic energy).

Rigidity (*load deflection rate* or *LDR*) of a material is proportional to the elastic limit, which determines the maximum (allowable working) load which the material can withstand. In a circular wire, the maximum load is proportional to the cube of its radius and inversely proportional to its length, while in a rectangular wire, it is directly proportional to the breadth of the wire and square of its height. The maximum deflection (maximum elastic strain) of a circular wire is inversely related to its radius and directly related to the square of its length, whereas in a rectangular wire, it is inversely proportional to its height.

Flexural rigidity is given by $(E.\pi.\ r^4)/4$, where E is the Young's modulus (of elasticity), and r is the radius of the wire.

LDR in the case of torsion of the wire is inversely proportional to the length of the torsion. For the rectangular wire, LDR is proportional to the cube of the magnitude of the wire in the direction of bending (flexure).

13.3 Properties of Biomaterials

The properties of biomaterials may be divided under the heads: mechanical (strain - tensile and compressive, stress–tensile and compressive, impact strength, fracture toughness, fatigue strength, hardness, abrasion resistance and time-dependent properties), rheological (viscosity, non-Newtonian flow and viscoelasticity), thermal (conductivity, diffusion and expansion), chemical and biological.

Tensile and compressive properties

Strain is the change in dimension (usually length) due to application of external forces or load. By applying *tension*, the length is increased, while by *compression*, the length is reduced. Strain is commonly expressed as $\Delta L/L$ and in percentage. It may be either *plastic* (permanent deformation) or *elastic* (temporary deformation).

The internal force, per unit area of a material, equal and opposite to the applied load or force is called *stress* and is measured as N/m^2. Stress may be compressive, tensile or shear. However, if a thick beam is bent with a 3-point loading, the lower portion of the beam will undergo tensile stress, while the top part will experience compressive stress. A slight shear stress may also be present.

The stress-strain relation is obtained by plotting stress against strain for a material in tension or compression. The representative graphs for a ductile metal, brittle ceramic and a polymer are as shown in Fig. 13.1. Here, *TS* denotes the tensile stress, *P* denotes the *proportional limit* (This is the maximum stress at which it is proportional to strain. Ceramics often break at this point), *E* represents E*lastic limit* (Often a material may be stressed above *P*, yet behave elastically. Elastic limit is the maximum stress tolerated by a material without permanent deformation) and *Y* is the y*ield point/strength* (The point where there is a rapid increase in strain without a corresponding increase in stress, leading to complete plastic deformation. Sometimes, e.g., in polymers, due to cross-sectional contraction or necking, there is an accompanying elongation with the volume remaining the same). Also please see Section 8.2.

Hooke's law is obeyed when stress is linearly proportional to strain. This linear slope reflects the rigidity of the material and is known as the *modulus of elasticity*. It is also known as *Young's modulus*, after its discoverer.

Young, along with Helmholtz, served both the physical sciences and the life sciences in many ways. They made great contributions in understanding the auditory and visual processing mechanisms in humans. Thomas Young is most famous for his work on light and vision. He showed that changes in curvature of the eye's lens focus light, found the cause of astigmatism, and proposed the correct three-colour explanation of colour vision. Young's experiments with interference of light waves and diffraction of shadows established the wave nature of light. He was the first to recognise that light travels as transverse waves. In other areas, Young was the first to define energy in the modern sense. He improved on Hooke's law by providing a measure – Young's modulus – showing the constant of proportionality between the force and stretching for different substances. He even helped establish the basis for translation of Egyptian hieroglyphics (of the Rosetta stone). In his lectures he put forward the hypothesis, afterwards developed by Helmholtz, that colour perception depends on the presence of three kinds of nerve fibres in the retina which respond respectively to red, green and blue light. Although Herman Ludwig Ferdinand von Helmholtz is known especially as a physicist and biologist, he was also a physician, mathematician, philosopher, and lecturer on popular science. He extended the application of the law of conservation of energy and in 1847 formulated it mathematically. He contributed to the knowledge of thermodynamics and electrodynamics and studied vortex motion in fluids. A pioneer in physiological optics and author of a Treatise on Physiological Optics (1867; tr., 3 vol., 1924-25), he extended Thomas Young's theory of colour vision, explained the mechanism of lens accommodation in the eye, and invented (1851) the ophthalmoscope. He was an authority on acoustics, especially on the perception of tone quality, and wrote *On the Sensations of Tone* (4th ed. 1877, tr. 1954). Helmholtz was professor of physics at the University of Berlin from 1871 and also director of the physicotechnical institute at Charlottenburg from 1887.

The Young–Helmholtz theory, developed in 1852 by Thomas Young and H. L. F. Helmholtz, is based on the assumption that there are three fundamental colour sensations – red, green, and blue – and that there are three different groups of cones in the retina, each group particularly sensitive to one of these three colours. Light from a red object, for example, stimulates the cones that are more sensitive to red than the other cones. Other colours (besides red, green, and blue) are seen when the cone cells are stimulated in different combinations. Only in recent years has conclusive evidence shown that the Young–Helmholtz theory is, indeed, accurate. The sensation of white is produced by the combination of the three primary colours, and black results from the absence of stimulation. In 1843 Ohm stated the fundamental principle of physiological acoustics, concerned with the way in which one hears combination tones. Helmholtz explained the origin of music on the basis of his fundamental physiological hypotheses and formulated a resonance theory of hearing which provided a physiological explanation of Ohm's principle. Young's modulus is usually denoted by E. Shear or elastic modulus (G) and bulk modulus (K) are also measured. The higher the modulus, the more is the rigidity and less the flexibility.

Due to practical difficulties, P and Y cannot be measured easily. The measuring instrument has to be very sensitive and accurate or it would affect the values. So, for the tensile testing of metals, proof stress values are used. 0.1% proof stress corresponds to 0.1% permanent (plastic) strain.

Drawing a metal in the form of a wire reflects *ductility*, while making it into thin sheets without breaks or cracks manifests its *malleability*. A very ductile metal will have a high percentage of elongation, whereas a highly malleable metal will show bigger reduction in cross-sectional area.

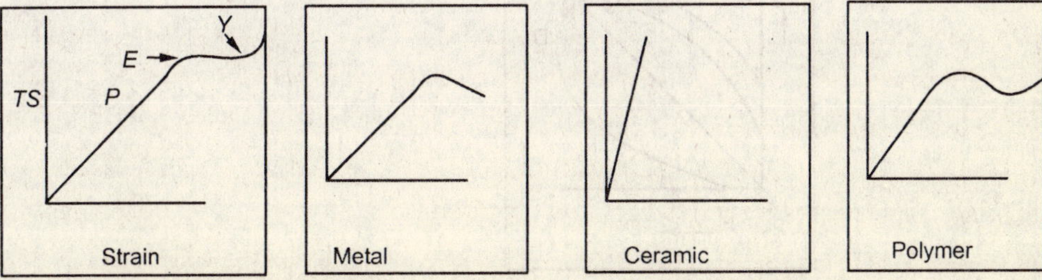

Fig. 13.1 Stress–strain relation

$$\text{Poisson's ratio or } \mu = \frac{\text{(lateral/transverse strain)}}{\text{axial strain}} = \frac{\text{(reduction in cross} - \text{sectional area)}}{\text{(increase in leangth) in the material under tension}}$$

Tensile strength or *ultimate tensile strength (UTS)*, is the maximum stress that a material can withstand without fracturing or the maximum resistance to fracture. It is usually calculated by the formula $2\rho/\pi td$, (where ρ denotes the load beyond which the material fractures, t is the thickness and d the diameter of the specimen) for a brittle material. The tensile strength is tested by the diametrical compression test, where a cylinder or disk is compressed through its diameter and tensile stress is set up at 90° to it. The material must follow Hooke's law up to the point of fracture for this formula to be valid.

Modulus of resilience is the maximum amount of energy, which a material can absorb before undergoing permanent deformity. It is denoted by $R = P/2E$, and is measured as energy/volume (J/m^3). *Toughness* of a material is its ability to absorb energy before fracture, and is also expressed in J/m^3.

In real life, instead of a gradual increase in load, there is often a sudden impact of great force leading to a break or fracture in the material. This can be determined by testing the impact strength with a Charpy tester (the specimen is supported horizontally at the ends) or Izod instrument (the material is clamped at one end and held vertically). *Fracture toughness* is the critical stress intensity factor for fracture propagation, usually denoted by K_{1c} (MN/m$^{1.5}$). *Fatigue* is caused by the repeated application of a small load. It is measured by bending or twisting a material through N cycles at a known tolerable stress S. S is plotted versus log N to determine the endurance limit below which fatigue fracture is not possible.

Hardness reflects the ability of the material to withstand indentation (with a known load for a fixed time). The depth and area of indentation are inversely related to hardness. *Abrasion resistance* is measured by scratching a material under test conditions and noting the depth of scratch or weight of material removed. The time dependent (viscoelastic) property of a polymer is due to relaxation with time and is reflected by the reduction of elastic modulus with time.

Rheological properties

Rheology is the science of studying the flow and deformation of a material. In the case of Newtonian (ideal) fluids, the plot of shear stress (T) versus shear rate (D) is linear, passing through the origin. In Fig. 13.2, N denotes the plot for a Newtonian fluid, D for a *dilatant* (e.g., wet sand with low water content), B denotes a *Bingham flow* (e.g., composite dental filling material), P represents a *plastic* material and a Ps plot may be obtained by a *pseudoplastic* fluid like blood, saliva, many polymers, natural resins, fluoride gels and viscous oils.

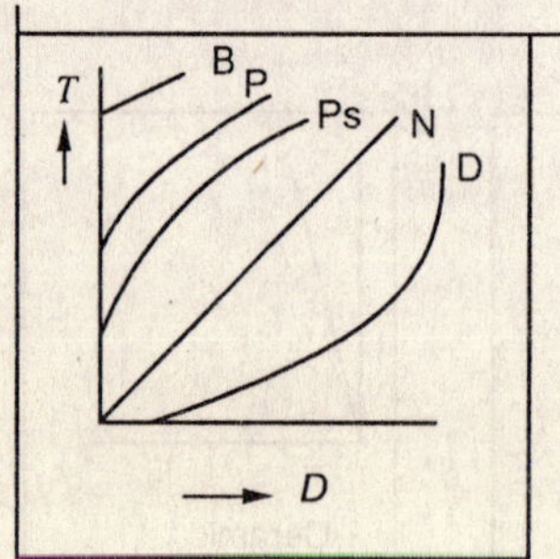

Fig. 13.2 Various types of fluid

Some other important properties included under the rheological classification are *working time* (period during which the material can be manipulated without thickening), and *setting time* (time when the material has reached a given degree of rigidity). *Viscosity* of a fluid is its resistance to flow. The *dynamic viscosity* unit is dyne s/cm^2, also known as Poise (in SI units, 1 Pa.s = 10 Poise), while the *kinematic viscosity* unit is m^2/s or m^{2}s^{-1}. The slope of the velocity–distance curve or shear rate $D = dv/dx$, while shear stress $\tau = F/A$ (dynes/cm^2), and the *coefficient of viscosity* is a constant $\eta = \tau/D$.

For viscous fluids, deformation is time dependent. The combination of viscous and elastic properties is called *viscoelasticity*, and has been discussed in relation to bones in Section 4.2.

13.4 Various Types of Biomaterials

Solid biomaterials may be divided into *polymers* (a long chain of repeating units of molecules), *metals* (these have good thermal and electrical conductivity properties), *ceramics* (compounds of metallic and non-metallic elements) and *composites* (a combination of more than one of these types e.g., polymer-ceramic). Among the various biomaterials, polymers are resilient and easy to fabricate, but not strong and deform with time. Metals are strong, tough, ductile and malleable, but may corrode easily, and are dense and difficult to make. Ceramics are very biocompatible, inert and strong in compression, but brittle, non-resilient and also difficult to manufacture. Composites are strong, can be tailor-made to suit any situation but difficult to prepare. A detailed comparison between the various materials is given in Table 13.1.

Table 13.1 Comparative characteristics of ceramic, metallic, polymeric and composite materials

Properties	Ceramic	Metallic	Polymeric	Composites
Chemical and rheological	Resistant to corrosion	Prone to corrosion	Viscoelastic	According to components
Mechanical	Stronger compression than tension; low impact strength and more brittle	Similar tensile and compressive strength; high impact strength and less brittle	Less rigid than ceramics and metals; capable of plastic deformation	Stiff, strong lightweight or resilient and compliant
Thermal	Insulator	Conductor	Addition polymers are thermosetting	May be insulator or conductor
Electrical	Insulator	Conductor	Insulator	Either
Stability at high temperature	Very high	Melts at very high temperature (much higher than attained in the human body)	High but less than ceramic	As in polymeric
Crystallisation	Less readily	Easily	Less readily	Less readily
Structure	Most complex, no free electrons leading to translucency	Simpler, free electrons leading to opacity	Depends on the condensation or addition of polymers (covalent bonds)	Shape, volume fraction and interfaces of hetero-geneities decide
Common uses	Dental & orthopedic	Dental & orthopedic	Orthopedic, suture, vascular, soft tissues	Joints and heart valves

Another commonly used biological (tissue derived) biomaterial is collagen (also see Section 7.3). It may be available as a film, a gel or a foam, which can be used either as stand-alone products or in combination with other products, e.g., by Imedex Biomatériaux

Collagen Film (COF)

This product has demonstrated anti-adhesion properties in multiple surgical settings. It is presently developed in combination with other medical devices with the objective of acting as a complementary anti-adhesion barrier. COF film has been associated with a three-dimensional hernial mesh from FLOREANE/SOFRADIM for abdominal wall repair under the trade name "Parietex Composite"®. It was tested in large clinical studies and was shown to be highly effective in preventing post-op adhesions of internal organs to the mesh.

FLOREANE/SOFRADIM has received a CE mark for this product, which is now commercialised in Europe and considered for introduction in the US market.

Collagen Dry Sealant (CDS)

This product is a bilayered material combining the COF film with a hemostatic collagen fleece. CDS is immediately and strongly adhesive. It can be applied to a wound to stop bleeding, while at the same time preventing post surgical adhesions.

Collagen Gel Sealant (GAO)

This fluid product is designed to rapidly form adhesive, elastic, fluid and air-tight seals on non-bleeding tissues while also providing anti-adhesive properties. It is ready to use in a simple kit within 3 minutes. It can be delivered directly or through a catheter for use in minimally invasive surgery. It can also be connected to a spray system.

GAO has been tested for multiple indications and has demonstrated excellent preclinical results which make it more attractive than other existing tissue sealants in the market.

Collagen Foam Sealant (MAC)

This product is also fluid and adhesive. It is prepared by the incorporation of air micro bubbles within the GAO gel sealant with a complementary kit, which makes it adapted for hemostasis of bleeding wounds. Its anti-adhesive properties are also being evaluated.

The products portfolio also includes different adhesive drug delivery systems developed from the three basic collagen biomaterials. Apart from these, a specific formulation of CDS is under development for association with cells; a specific formulation of GAO is under development for cartilage and bone repair and a specific formulation of MAC is being developed in association with antibiotics or growth factors for healing chronic wounds of the skin.

Bioplastics

Plastics contain one or more polymers and are solid in the finished stage. Most of these polymers are synthetic and come from non-renewable natural resources like petroleum, natural gas and coal. Biopolymers like carbohydrates and proteins also occur naturally. Starch-based bioplastics are the least expensive and can be processed by all the methods used for polymer synthesis. Starch-protein compositions can meet nutritional requirements of farm animals.

Collagen (Section 11.3) is the most abundant mammalian protein. When it is denatured, it becomes gelatin which can be used in sausage casing, capsules for drugs and also in photography. Polyesters are produced by bacterial fermentation in large scale. Now with natural resources like starch and sugar, water-resistant bottles are being manufactured. Lactic acid, which is produced commercially by fermenting sugars, can be polymerised to form polylactic acid that is utilised in drug encapsulation and biodegradable medical devices. Triglycerides can also be polymerised.

Bioplastics are biodegradable and are derived (almost) entirely from renewable raw materials. Celluloid (invented by John Wesley Hyatt, Jr., in the 19th century) was the first bioplastic. It was derived from plant cellulose.

Commercially available bioplastics are generally dearer than synthetic polymers. Only starch-based polymers can now compete with them. A combination of petrochemical synthetic polymers like polyethylene or polystyrene and plant polymers allows a better microbial degradation of synthetic polymers. Usually starch is used in a native or modified form, or along with synthetic copolymers (ethylene/acrylic acid, polyvinyl alcohol or polystyrene chain grafted on amylose or amylopectin chains). The starch in it confers opacity, water-sensitivity and humidity that may restrict their uses. Bacterial polyesters like PHB (poly hydroxy butyrate), PHA (poly hydroxy alkanoate), PHV (poly hydroxyalkanoate valerate) and PLA (poly lactic acids) are totally biodegradable, and not bio-fragmentable like the starch–petroleum combinations.

To reduce the cost of bioplastics production, it may be coupled with conventional (bacterial) wastewater management system. Another approach is to produce "green" plastic (PLA or PHA) in plants through bacteria like *Ralstonia eutropha* and *Escherichia coli* by genetic engineering or by using biomass derived renewable energy.

13.5 Biocompatibility

Biocompatibility is nothing but the ability of a foreign body to coexist in a living (whole body in general and particularly surrounding) tissue. Whether a biomaterial will succeed inside the body, depends on the material's properties, its design, on the surgical technique used, and also on the physical condition and activity level of the host (recipient of the implant). One of the reasons for failure could be host reaction to the implant (foreign body) leading to *implant rejection*, while the other end of the spectrum could include implant- induced toxicity, allergy or even malignancy (cancer) to the host. If, instead of rejection of the graft by the host, the graft itself causes severe damage to the recipient, it may be termed as *graft versus host reaction (GVHR)*.

Biocompatible materials are NOT supposed to irritate, inflame, and cause allergy or cancer. Other important compatibility characteristics of implant materials are mechanical (strength or working capacity, stiffness or fracture resistance and fatigue resistance) and optical (for eye, skin or teeth) properties, adequate density, sterilisability, manufacturability, long-term storage and proper engineering design.

Assessment for biocompatibility is done by systemic toxicity study, skin sensitisation/ irritation, tissue/bone implantation tests on animals like albino rats, rabbits, guinea pigs, hamsters; in-use tests in monkeys/dogs and cytotoxicity tests *in vitro* (cell culture).

Failure may be time dependent e.g., infections which occur immediately, and implant fracture which occurs later. Also an artificial blood vessel graft may clot much earlier than a natural vessel.

To quantify an abstract thing like reliability (r) of the implant, in terms of the probability (f) of implant failure, $r = 1 - f$. In the case of multiple modes of failure, the total reliability $r_t = r_1 . r_2 . . . r_n$.

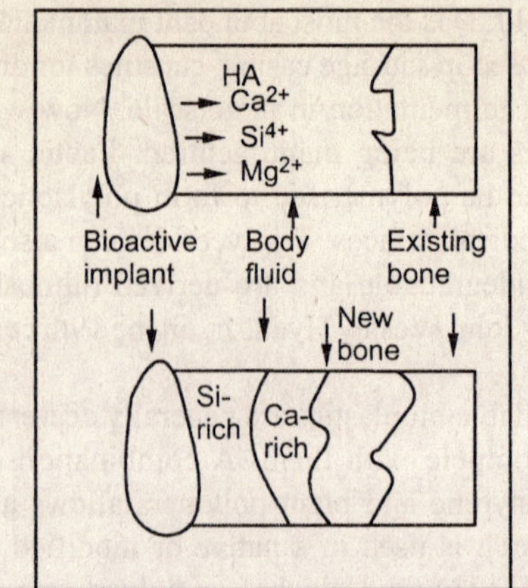

Fig. 13.3 Reaction of body tissues with a bioactive implant

The implant may not always be biologically inert (either intentionally or unintentionally). To cite an example, the mechanism of new bone formation and bone bonding to a bioactive (biodegradable or resorbable) ceramic implant is illustrated in Figure 13.3. Immediately following implantation, an ion-exchange reaction takes place between the implant and the surrounding body fluid during which chemical species from the ceramic diffuse into the fluid and vice versa. Over time, this results in the formation of chemically graded layers that become hydrocarbonate apatite, or new bone.

This characteristic property may also be utilised as drug delivery devices, for filling spaces vacated by bone screws, donor bones or excised tumors, repairing and fusing spinal vertebra, repairing herniated disks and repairing maxillofacial and dental deficits.

13.6 Further Reading

[1] Black, J. *Biological Performance of Materials.* 2nd ed . NY: M Dekker Inc. 1992.

[2] Greco, R. S. *Implantation Biology.* Boca Raton, FL : CRC Press. 1994.

[3] Van Vlack, L. H. *Elements of Materials in Science and Engineering.* 5th ed. Reading, MA: Addison Wesley. 1985.

13.7 Questions

1. Descriptive Questions

1. What is a biomaterial and what are its applications ?
2. Which biomaterials are commonly used for cardiac prostheses ?

2. Write Short Notes on

1. Biocompatibility
2. Biomaterial failure
3. Biopolymers

4. Ceramic biomaterials
5. Composite biomaterials

3. Multiple Choice Questions

Choose the single best response.

1. A biomaterial can be
   (A) a drug
   (B) synthetic or natural
   (C) either
   (D) none
2. The success of a biomaterial is dependent on all except
   (A) physical condition of the host
   (B) design
   (C) surgical technique
   (D) cost
3. Which is not a mechanical property of a biomaterial ?
   (A) viscoelasticity
   (B) strain
   (C) fracture toughness
   (D) abrasion resistance
4. Which type of fluid is blood ?
   (A) Newtonian
   (B) dilatant
   (C) pseudoplastic
   (D) Bingham
5. Bioplastics are
   (A) cheap
   (B) biodegradable
   (C) commonly used
   (D) liquid

Answers

1. B; 2. D; 3. A; 4. C; 5. B

Section 5

Kinesiology and Biological Control

Chapter 14 deals with kinesiology of human musculoskeletal system, orthopedic biomechanics, and locomotion analysis.

The biomechanics of human motion and control interfaces with application to limb orthotics and prosthetics are described in Chapter 15.

Principles of feedback control, fluid (blood/air) mechanics in biology, neuromuscular control systems and modelling of those are discussed in Chapter 16.

14

Kinesiology of Human Musculoskeletal System, Orthopedic Biomechanics, Locomotion Analysis

14.1 Introduction

The study of different types of muscles, lever systems and their movements is called *kinesiology*. It is the sum of what is known regarding the motion in the human body. Muscles usually operate by applying tension to their points of attachments to bones, which in turn, act as levers. An analysis of lever systems of the body depends on

(i) a definite knowledge of the point of muscle attachment,
(ii) its distance from the fulcrum of the lever,
(iii) length of the lever arm and also
(iv) position of the lever.

Our body performs various types of movements – some needing much strength and others which require moving through great distances. So, we have different types of muscles – some are long and contract over a large distance, while some are short but have large cross-sectional areas, thus purveying great strength of contraction over short distances. If a bone is broken, and then heals in a shortened state, the contraction force of the muscle lying along this broken bone would obviously decrease due to the shortened length of muscle. These muscles undergo compensatory physical shortening during the next few weeks after a fracture and establish new muscle length, which is approximately equal to the maximal length of the lever system itself – thereby re-establishing optimum force of contraction by the muscle. Similar shortening occurs in muscles of limbs immobilised for many weeks in casts (in case the muscles are in a shortened position during that time). After removal of the cast, the muscle must be restretched for weeks to restore full mobility.

Kinematics describes spatial motions (angular displacements) of joints and limb segments. Qualitative kinematics is observational, while kinetics (quantitative) describes moments or torques and forces that cause joint and limb motion. Firing patterns of muscle can be measured with dynamic electromyography (EMG). Please see Section 2.4.

Kinetics includes the study of muscular activity and the forces acting there. There are reactive joint moments produced around the lower limb joints by active and passive soft tissue forces and associated mechanical power of the joint moment. A moment around a joint occurs when a force is acting at a distance from the joint, through a lever, accelerating the joint angle. For example, an externally applied extensor movement about the elbow is produced when a weight is placed in the hand – where the lever is the forearm (radius and ulna) and the elbow will tend to accelerate into extension. External moment is calculated as $W \times L$ (where W is the weight of the object and L is the length of the forearm). To keep the joint angle stable, all moments acting around the joint must add up to zero. An internal force from the biceps acting through its forearm lever can purvey this resisting internal flexor moment, stabilising the elbow joint. In accordance with the magnitude of force through the biceps, the elbow joint angle will extend in a controlled fashion (eccentric contraction), stay as it is (isometric contraction) or flex (concentric contraction). Externally applied moments from gravity, inertia and the ground are balanced by internal joint moments from muscular and/or soft tissue activity.

Ergonomics is the discipline that investigates and applies information about human requirements, characteristics, abilities and limitations, to the design, development and testing of equipment, systems and jobs. It attempts to improve performance by optimising the relation between people and their workplace. Also please see Section 10.7.

14.2 Bones as Lever Systems

Orthopedic biomechanics deals with the mechanics of bone and the adjoining musculoskeletal system during various activities, like lying down, bending, standing, sitting, squatting, walking and running (various *static* and *dynamic postures*). However, a detailed kinesiological analysis of each of the postures is beyond the scope of this book.

The long bones form a light framework of levers. The skeletal muscles attached to them contract to operate/move these levers. Muscle contraction leads to shortening, in other words, bringing the two ends of attachment closer. The two ends are attached to different bones by different tendons, and any one bone will move. Two or more bones articulate at a joint – whose surfaces are covered with a layer of smooth (hyaline: see Section 11.2) cartilage and a synovial membrane. The synovial membrane secretes a synovial fluid (lubricant) to avoid friction between movements of two surfaces of bones.

The main mover or *agonist* or *protagonist* (e.g., biceps brachii for flexing or bending the elbow) contracts to move the joint. However, for smooth coordinated movement, the opposing muscle (e.g., triceps brachii for flexing the elbow) or *antagonist* must relax. Again, for the opposite movement, i.e., extending or straightening the elbow, the triceps becomes the agonist, while the biceps becomes the antagonist. The mechanism involved in this is explained in the next section). Some other muscles are also involved during movement of a joint. Fixators steady the bone with one attachment (formerly called the 'origin') of the agonist muscle so that only the other attachment (earlier known as 'insertion') will move. Synergists steady the joint being moved. The relation between muscle length and tension is discussed in the next section.

Mobilisation and strengthening of stiff knee/shoulder, or other joints

Movements around a joint may be active, passive or abnormal. For assessing normal movement, clinical acumen is sufficient. Rarely is the goniometer (Section 16.4) required. The range of joint movement starts from 0° (neutral or anatomical position of the joint). Flexion and extension (bending/straightening) occurs in the sagittal plane (*see* Appendix 1), as in the knee, elbow, ankle, fingers and toes. Adduction and abduction (bringing in/out) take place in the coronal/frontal plane, toward or away from the midline. Pronation and supination are the rotatory movements of the forearm and the foot. Circumduction is a composite movement (rhythmic sequence of the other movements) in the ball and socket joints like hip and shoulder. The thumbs can have a further movement called opposition, which is so essential and yet so difficult to imitate in the case of amputation of the thumb, and another finger from the same hand is transplanted in the place of the thumb (pollicisation operation in plastic surgery). The spine can have lateral flexion and rotation movements, while the feet can undergo eversion and inversion. Following a disease or trauma, joint stiffness is measured in terms of loss of all movements, limitation of all movements and limitation of one or two movements. Joint laxity may also have to be tested in some conditions.

Often, due to some disease or degeneration, a particular joint e.g., the shoulder joint may become stiff (commonly known as frozen shoulder). The aim of the therapists should be to restore normal movements in the joint, as far as practicable. For this, various methods are used (the most common method is to apply ultrasound waves to the stiff joints). The details has been dealt with in Section 2.11. Occasionally, other types of therapy (pharmacotherapy or medicine) like intra articular (within joint capsule) injection of corticosteroids are also necessary. However, all these types of therapy can become successful if and only if a judicious concomitant regimen of physical (exercise) therapy is followed strictly.

14.3 Muscle Tone

Some idea about muscle contraction has already been purveyed in Section 10.3.

Tone of a muscle is defined as the state of partial sustained contraction during rest. The alternative definition for *tonus* or tone is the resistance of a muscle to stretch. It is maintained as a *reflex* (spontaneous response not directly involving higher centres of the central nervous system) and therefore requires an intact synaptic connection. As a reflex requires an afferent (sensory) neuron, efferent (motor) neuron and an intact spinal centre, if there is a lesion anywhere in the whole circuitry, the reflex will be lost or reduced (*hypotonia*). Moreover, there are inhibitory influences from the higher centres in the brain and brainstem (almost like a parental restriction to a child's freedom) over the reflexes. So, if there is a lesion in the higher centres (upper motor neuron), the reflex maintaining the tone will be exaggerated (hypertonia or spasticity), as in the cases of stroke or cerebrovascular accidents (CVA). On the other hand, lesions of the lower motor neuron (anterior horn cells of spinal cord or below) will lead to hypotonia or flaccidity. Similarly, in the cases of sensory (afferent) defects too, there will be hypotonia. Muscle tone is usually dependent on the discharges from the g efferent motoneurons. Disorders of the cerebellum often lead to hypotonia. Another condition where the muscle tone is reduced is the paradoxical sleep or rapid eye movement (REM) sleep, when we mostly dream.

However, for understanding muscle tone better, we must know learn about the structure of muscle spindles (stretch receptors). Each muscle spindle comprises about ten muscle fibres enclosed in a connective tissue capsule, and are known as *intrafusal fibres*. The regular contractile units of the muscle (thick and thin filaments) are called the *extrafusal fibres*. The intrafusal fibres are in parallel with the other fibres, as the ends of the spindle capsule are attached to the tendons at both ends of the muscle or to the side of the extrafusal fibres.

Intrafusal fibres may be further subdivided into a centrally dilated nuclear bag fibre (2 per spindle) and thinner nuclear chain fibre (4 or more per spindle). Similarly, sensory endings are also of two kinds, viz., *primary* or *annulospiral* endings, where rapidly conducting group Ia afferent (sensory neural) fibres end, and the *secondary* or *flower-spray* endings, where group II afferent fibres terminate on the nuclear chain fibres. Of the Ia fibres, one branch innervates nuclear bag fibre 1 (low level of myosin-ATPase activity), while another branch innervates nuclear bag fibre 2 (high level of myosin-ATPase activity) and nuclear chain fibres, wrapping around their centres. The muscle spindles have exclusive motor supply from the small Aγ fibres or γ efferents (also known as gamma motoneurons). Additionally, larger β motoneurons innervate both intrafusal and extrafusal fibres. Both sensory and motor neurons produce dynamic and static responses. Unlike the dynamic fusimotor axons, stimulation of the static fusiform axons increases static discharge at constant length. The nerves ending in nuclear bag fibres discharge most rapidly while the muscle is being stretched, and less rapidly during sustained stretch (dynamic response). On the other hand, nerves terminating on nuclear chain fibres discharge at a high rate throughout the period when a muscle is stretched (static response).

Another important concept, related to stretch reflex (or deep tendon reflex, like a knee jerk), is called *reciprocal innervation*. This arrangement makes the antagonist (opposing) muscle to relax, when a particular muscle (agonist or protagonist) is contracting, as described in the previous section. Impulses in the Ia fibres from the spindles of the agonist muscle cause postsynaptic inhibition of the antagonist motoneruons. This is a bisynaptic reflex. A collateral from each Ia fibre passes in the spinal cord to an inhibitory (Golgi bottle neuron) interneuron that synapses directly on one of the motoneurons supplying the antagonist muscles.

Receptors for sensing pressure changes in muscles are termed *Golgi tendon organs (GTO)*. Unlike the spindles, they are in series with the muscle fibres, and are therefore stimulated by both passive stretch and active contraction of the muscle. It has a low stimulation threshold. The GTO functions as a force transducer in a feedback circuit regulating the muscle force, similar to the muscle spindle feedback circuitry regulating the length of the muscle. It mediates the inverse stretch reflex or autogenic inhibition, where when a muscle is stretched to a great extent, the muscle suddenly relaxes, instead of contracting further.

The total tension developed by a skeletal muscle during isometric contraction and the passive tension exerted by the unstimulated (resting) muscle vary with the initial length of the muscle fibre. The difference between the two is known as active tension. The length of the muscle at which the active tension is maximal is commonly known as the *resting length* since it has been experimentally found that the length of most muscles at rest is that at which they develop maximal tension. In accordance with the sliding filament theory (Section 10.3), in isometric contraction, the tension developed in the muscle fibre is proportionate to the number of cross-linkages between the actin and the myosin molecules. When a muscle stretches, the actin-myosin overlap and the number of cross-linkages decrease. Again, when the muscle is shorter (contracted), the thin filaments overlap, decreasing the number of cross-linkages.

The velocity of muscle contraction also varies inversely with the load on the muscle. At a particular load, the maximal velocity is obtained at the resting length only.

14.4 Locomotion (Gait) and its Analysis

Gait or *locomotion analysis* is the quantitative measurement and assessment of human locomotion including walking and running. It is usually done bare-foot, but may be done even with lower limb orthoses or prostheses.

In the case of human locomotion, momentum and balance are integrated into body movement in such a way that maximum motion is possible with minimal muscle exertion. So, the stress on tendons and bones rarely exceed 50% of their failure strength. To cite an example, the hamstring muscles, extending from the pelvis over the hip joint and knee joint, on contraction, produce flexion of leg at the thigh. If the thigh is flexed on the pelvis simultaneously, lengthening of the hamstrings across the hip joint compensates for their shortening across the knee joint.

Postural adjustments and voluntary movements are very much similar and interrelated. A major factor in postural control is the variation in threshold of spinal stretch reflexes that is caused by alterations in the excitability of the motoneurons, and also indirectly by changes in the γ efferent discharge rate to muscle spindles.

For locomotion (walking) the four essential steps are:

(i) increased tone of anti-gravity (extensor) muscles of the lower limbs,
(ii) stepping,
(iii) control of centre of gravity to purvey equilibrium, and
(iv) a mechanism for acquiring forward motion.

These steps must be very well coordinated for an effective smooth walking movement. For maintaining the equilibrium, five intact systems are necessary, viz., visual, labyrinthine (utricles, saccules and semicircular canals in the ear), limb receptors (in muscles, joints and tendons), pressure receptors (in toe pads and other parts of the body) and stretch receptors (in muscles, joints and ligaments of the neck). The stretch and pressure receptors of the muscle have already been discussed in the previous section. The single most important factor, however, is the distribution of muscle tone.

Important reflexes required for maintenance of posture and equilibrium are:

(1) local segmental static reactions, e.g.
  (i) positive supporting reaction,
  (ii) negative supporting reaction,
  (iii) segmental static reaction;

(2) general static reactions, e.g.
  (i) statotonic or attitudinal reflexes, like
    (a) tonic labyrinthine and tonic neck reflexes acting on limbs,
    (b) tonic labyrinthine and cervical reflexes acting upon eyes and
  (ii) righting reflexes;

(3) kinetic, phasic, statokinetic or acceleratory reactions either
  (i) angular (reaction to rotation of head and eyes) or
  (ii) linear (progressive movement of head and extremities) like natural righting response. These are all short-term (transient, not sustained) reflexes.

While walking, the rhythmic stance (support) phase or period (foot on ground) alternates with the swing phase or period (foot off ground). The support phase of the legs overlap, so, there are two periods in each cycle when the body is supported by both legs. There is a brief burst of activity in the leg flexors at the beginning of each step and the leg is swung forward with little more active muscle contraction. That means, muscles are active for only a fraction of each step and therefore, long walks usually do not cause fatigue. For a young adult, about 80 m/min is a comfortable pace with optimum use of energy and with 150–175 W power generated per step.

Gait being a cyclic activity, the *gait cycle* can be defined as the period of time from the point of initial contact (foot contact) of the subject's foot with the ground, to the next point of initial contact for the same limb. The stance and the swing phases are demarcated at the point when stance leaves the ground (*toe off* or *foot off*).

There are eight phases of the gait cycle.

1.	Initial contact	Stance period →	Tasks: Weight acceptance and Single limb support.
2.	Loading response		
3.	Mid stance		
4.	Terminal stance		
5.	Pre swing	Swing period →	Task: Limb advancement.
6.	Initial swing		
7.	Mid swing		
8.	Terminal swing		

During swing, most external moments occurring around the lower limb joints are the result of gravity and the inertial forces from the individual limb segments. For instance, during swing, both weight of foot and inertial force from the leg swing generates external plantar flexion that should be restrained by internal dorsiflexion by the ankle dorsiflexors to avoid footdrop.

During the stance phase, most external moments are produced by the ground reaction force (GRF). While standing quietly without movement, the body weight on the ground is pushed by equal and opposite GRF (forces and torque), whose vector passes through the base of support or feet up toward the centre of mass (COM) of the body. While walking, GRF is overcome by body weight plus acceleration/deceleration with the up/down movement of the COM.

Commonly gait analysis requires the following data

1. Static physical measures like passive joint range of motion (Section 16.2), muscle strength (Section 10.3) and tone (Section 16.3), and the presence and degree of bony deformity, if any
2. Stride and temporal parameters, like step length and walking speed
3. Kinematics (Section 16.1)
4. GRF
5. Kinetics (Section 16.1).
6. Measurement of muscle activity during gait or dynamic EMG (Section 2.4)
7. Measures for metabolic energy expenditure (Section 10.6)
8. Videotape of the gait trial for qualitative review and control.

However, a lot of precautionary measures must be taken to obtain the above data. Some of the points in a standard data collection protocol are:

1. The patient must be made to feel comfortable.
2. The methodology and the equipment used must not alter the patient's gait.
3. Preparation and testing time must be minimised, including rest and play intervals.
4. Data collection techniques must be reproducible.
5. Robust yet flexible methodology is essential to evaluate all sorts of deviation from normality.
6. Validation (processing) of the raw data collected must be done before the patient leaves the test centre. Software packages like SIMM or Gait Lab may be used.

For motion measurement, electrogoniometry (with a rotary potentiometer having arms fixed to the shaft and base for attachment to the extremity across the concerned joint) or multi-axial goniometry (with real time display and rapid collection of single-joint information on many patients), and high speed photography (cinefilm), especially for running may be used. Other methods include accelerometry (multi-axis, both linear and angular) and video filming.

For GRF measurement, force plates or platforms may be used. The three (X,Y,Z) components of the GRF vector, ground reaction torque (vertical) and the point of application of the GRF vector (centre of pressure) are measured with these platforms embedded in the walkway. These plates of 0.5×0.5 m size, consist of several strain gauges or piezoelectric sensor arrays mounted together and held tightly.

For collection of the data for gait analysis, several external markers have to be placed on the surface of body segments that are aligned with particular bony landmarks. These markers are usually attached to the patients as discrete units or in rigidly connected clusters.

However, gait analysis does have its limitations. They are due to sources of error like soft tissue displacement relative to bone, estimates of joint centre locations, especially the hip, approximations of the inertia of body segments, and also the numerical differentiation of noisy displacement data. Other minor errors are due to faulty data collection, like improperly placed markers or inadvertent contact of a force platform with the swing limb. Patients with lower body surface area (e.g., children) have lesser intermarker distances, therefore, angular computations become imprecise.

14.5 Further Reading

[1] Winter, D.A. *Biomechanics and Motor Control of Human Movement.* 2nd ed. New York: Wiley. 1990.

[2] Schmidt, R. A., ed. *Motor Control and Learning.* 2nd ed. Champaign, Illinois: Human Kinetics Publishers. 1988.

[3] Whiting, H. T. A., ed. *Human Motor Actions: Bernstein reassessed.* Amsterdam: Elsevier. 1984.

[4] Rose, J. and Gamble, J. G. *Human Walking.* 2nd ed. Baltimore: Williams & Wilkins. 1994.

[5] Allard, P., Stokes, I. A. F. and Blanchi, J. P., eds. *Three-dimensional Analysis of Human Movement.* Champaign, Illinois: Human Kinetics. 1995.

[6] Harris, G. F. and Smith, P. A., eds. *Human Motion Analysis.* Piscataway, NJ : IEEE Press. 1996.

14.6 Questions

1. Descriptive Questions

1. What do you understand by the term 'kinematics' ?
2. How to do locomotion or gait analysis ?

2. Write Short Notes on

1. Muscle attachments
2. Anti-gravity (extensor) muscles
3. Centre of mass
4. Labyrinthine system for equilibrium
5. Limb receptors for pressure and stretch

3. Multiple Choice Questions

Choose the single best response.

1. The study of different types of muscles, lever systems and their movements is called:
   (A) kinesiology
   (B) kinetics
   (C) kinematics
   (D) ergonomics

2. The locomotor system of the human body uses as lever
   (A) muscles
   (B) bones
   (C) tendons
   (D) none of these

3. Muscle tone is low in
   (A) lower motor neuron palsy
   (B) cerebellar disorders
   (C) REM sleep
   (D) all

4. Reciprocal innervation is achieved by a neural pathway that is
   (A) monosynaptic
   (B) bisynaptic
   (C) polysynaptic
   (D) asynaptic

5. Which phase of gait cycle is not related to limb advancement ?
   (A) initial swing
   (B) mid swing
   (C) mid stance
   (D) terminal swing

Answers

1. A; 2. B; 3. D; 4. B; 5. C

15

Biomechanics of Human Motion and Control Interfaces with Application to Limb Orthotics and Prosthetics

15.1 Introduction

Cybernetics is the science of the processes of communication and control in animal and machine. It may also be called the study of the theory of control systems, especially the comparison of the machine and biological nervous system.

The term itself originated in 1947 when Norbert Wiener used it to describe a discipline which was apart from, but which touched upon, such established disciplines as electrical engineering, mathematics, biology, neurophysiology, anthropology, and psychology. Wiener, Arturo Rosenblueth and Julian Bigelow needed a new word to refer to their new concept, and they adapted a Greek word meaning 'steersman' to invoke the rich interaction of goals, predictions, actions, feedback and response in systems of all kinds (the term 'governor' derives from the same root) [Wiener 1948]. Early applications in the control of physical systems (aiming artillery, designing electrical circuits and maneuvering simple robots) clarified the fundamental roles of these concepts in engineering; but the relevance to social systems and the 'softer' sciences was also clear from the start.

In working with abstract concepts common to all systems, early cybernetic researchers quickly realised that the 'science of observed systems' cannot be divorced from the 'science of observing systems' [von Foerster 1974] – because it is we who observe. The cybernetic approach is centrally concerned with this unavoidable limit of what we can know: our own subjectivity. In this way cybernetics is aptly called 'applied epistemology'. At the least, its utility is the production of useful descriptions, and, specifically, descriptions that include the observer in the description. Cybernetic descriptions of psychology, language, arts, performance, or intelligence (to name a few) may be quite different from more conventional, hard 'scientific' views – although cybernetics can be rigorous too. Implementation may then follow in software and/or hardware, or in the design of social, managerial and other classes of interpersonal systems.

Let us now understand the processes involved in limb movement and building implants. To build an implant (e.g., limb orthotics and prosthetics – Chapter 12), the common steps are as follows: Initially, it would be ideal to prepare a model keeping in mind the general specifications and constraints involved in a given particular circumstance. If the model is well-specified, the next step would be optimisation. Otherwise, (for ill-drawn specifications) there is the need for re-evaluation. After this, the model has to be validated (if possible first *in vitro,* then *in vivo*, and in some cases in other animals too). Finally, there is the necessity for individualisation.

Models of the control of human movement (especially of the extremities) compactly summarise known system properties like electrical, chemical and mechanical phenomena. Quantification leads to a better understanding of the dynamics in normal and abnormal (diseased) conditions and can often predict some diagnostic and prognostic features. Assuming limbs to be rigid bodies, well-known laws of mechanics like D'Alembert's law, Newton – Euler principle and LaGrange equations may be used.

Analytic limb models can be described by sets of ordinary differential equations (ODE). In non-analytic models, expert systems, fuzzy logic and other tools of soft computing can be used. Then there are hybrid controller models which combine both analytic and non-analytic techniques.

Control of human movements is very complex. It deals with a large-scale, non-linear, time-variable, dynamically non-determined, redundant system. In all biological systems there is redundancy. If we had to optimise (in technological terms) the biological system, a far less amount of subsytems and components would have sufficed. However, the beauty (of efficiency) of all biological systems lie in the fact that this redundancy takes care of many unforeseen circumstances and is therefore far more robust (adaptable) than many optimised technical systems artificially constructed for the same job. As a dynamical system may change its states continuously, without an optimisation criterion, all trajectories of it are equivalent. Analytically, controllers can be simulated with the help of mechanical and mathematical tools. If we are preparing a non-analytic controller, the operation is modelled on a set of mappings between the inputs and outputs, similar to biological control, instead of the system,.

Within the general older population, general impairments contribute more substantially to walking disability than do common chronic diseases. The diagnosed diseases do not explain the impairments that led to the walking disability. Therefore, especially in this category of the population, clinicians should focus not merely on common chronic diseases but particularly on general impairments as targets for diagnostic analysis and treatment to decrease the walking disability.

15.2 Modelling for Implants

Before getting into the details of modelling, we should know that harnessing the upper limb amputee can be challenging when confronted with specialised needs. Often standard harnessing, though adequate, has limitations when we are trying to optimise prosthetic performance. A basic understanding of biomechanic worksources, harnessing objectives, and forces diagrams are essential for developing control harness variations. In most cases starting with a basic harness and adding features in a logical progression works best, rather than ad hoc 'innovation'.

Harnessing Objectives: A control harness has four main objectives: force transmission, suspension, comfort, and cosmesis.

Force transmission refers to the cable function connected to the biomechanic worksources through the harness. Friction must be minimised for effortless control. This is done by creating an optimal line of pull along with friction reducing components such as teflon liners and cable alternatives (Spectra). The worksources chosen must be movements that are both inconspicuous and independent of one another during operation.

Good suspension starts by analysing the types and degree of forces the harness is to control such as axial loading, transverse rotation, and longitudinal moment. The prosthesis must be stable in all normal positions of operation and resist the tendency to migrate. Axial loading requires distribution of the forces over as wide a surface area as possible without compromising the donning ease or cosmesis. Good suspension is also greatly dependent on the patient's body type and posture. Amputees may or may not be candidates for other harness variations because of poor posture, load intolerance, or other disabilities.

Comfort and cosmesis are equally important because both, though non-functional, greatly impact the ultimate usage of the prosthesis. Donning ease should also be mentioned along with comfort. A harness may provide good suspension, but may be difficult to don and doff correctly. New materials are now used which increase harness comfort and durability. The harness itself must also have a professional appearance. Buckles must be minimised, straps must lie flat, and the overall harness bulk must be reduced.

Axillary versus elbow crutches

Restoration of normal walking is one of the goals of management of most orthopedic disorders of the legs and spine. Many aids are prescribed for retraining an individual to walk normally in the early stages of recovering from fractures, operations or paralysis of the legs. The common appliances used are walkers, crutches and sticks which help the patient to put part of the weight on the upper limbs while walking.

Walking frames are used by elderly patients who are unsteady in gait and suffer from post-operative pain in the hip or knee. It comprises four tubular uprights joined on three sides by upper and lower horizontal tubes. One side is open for the patient to enter and stand surrounded by the frame on three sides. Hand grips are added on the upper horizontal rods. The lower tips of the four uprights are covered by rubber bushes to prevent slipping. The patient stands inside the frame, lifts the frame by holding the hand grips, moves it forward and rests it one step ahead. He then takes a step, lifts and takes the frame forward again. This is repeated and the patient progresses. The frame is used mainly indoors.

Crutches help in non-weight bearing walking when one leg cannot be rested on the ground due to pain or the presence of a plaster cast. The modern *axillary crutch* consists of two wooden uprights connected at the top by a padded transverse bar to fit into the arm pit. The body weight is borne through a transverse hand grip fixed at an appropriate distance lower down. The position of the hand grip should be such as to keep the elbow in 30-degree flexion. The two uprights are fixed to a single upright to rest on the ground. The lower end is rubber-tipped. The *elbow crutch* is a modified walking stick which provides in addition to the hand grip an additional support to the forearm. There is an upward extension, set a little obliquely, with a cuff attached to the upper part of the extension. The cuff, made of thin metal sheet, partly encircles the upper forearm and opens superiorly for the forearm and is lined with soft leather. The weight is borne partly through the hand and partly by the elbow pressing at the cuff. Hence the name elbow crutch. Elbow crutches can also be made of hollow aluminum tubes, where the height is adjustable by telescoping the tubes, which snap-lock through perforations at the required length.

A *walking stick* is prescribed to patients who need partial weight bearing on the hip or knee in case of pain in these joints. A walking stick also improves stability in gait in elderly patients. The length should be from the top of the greater trochanter to the ground. It should be such that the elbow is at 30-degree flexion when the patient puts weight on the hand grip. The tip should be rubber-tipped to prevent slipping. The tip is usually single but could end as a tripod or quadripod to improve stability.

An example of modelling is presented by Agarwal et al. (1984). The dynamic response of the human ankle joint to a band-limited random torque perturbation superimposed on a constant bias torque was observed in normal human subjects. The applied torque input, the joint angular rotation output, and the electromyographic activity were recorded using surface electrodes on the extensor and flexor muscles of the ankle joint. Transfer function models using time series techniques were developed for the torque (observed by the angular rotation of the input-output pair) and for the angular rotation (observed by the electromyographic activity of the input-output pair). A parameter constraining technique was applied to develop more reliable models. It is shown that the asymptotic behaviour of the system must be taken into account during parameter optimisation to develop better predictive models.

15.3 Modelling for Movements (Limbs)

As mentioned before, models of control of human movement (especially of the extremities) compactly summarise known system properties like electrical, chemical and mechanical phenomena. First let us refresh (Section 10.2.3) our ideas of the mechanics of muscle work.

Mechanics of muscle work

In studying the mechanical action of muscles, the individual muscle cannot always be treated as a single unit, since different parts of the same muscle may have entirely different actions, as with the pectoralis major, the deltoid, and the trapezius where nerve impulses control and stimulate different portions of the muscle in succession or at different times. Most muscles are, however, in a mechanical sense, units. But in either case, the muscle fibres constitute the elementary motor elements.

Direction of muscle pull

In those muscles where the fibres always run in a straight line from the origin to insertion in all positions of the joint, a straight line joining the middle of the surface of the origin with the middle of the insertion surface will give the direction of the pull. If, however, the muscle or its tendon is bent out of a straight line by a bony process or ligament so that it runs over a pulley-like arrangement, the direction of the muscle pull is naturally bent out of line. The direction of the pull in such cases is from the middle point of insertion to the middle point of the pulley where the muscle or tendon is bent. Muscles or tendons of muscles which pass over more than one joint and pass through more than one pulley may be resolved, so far as the direction of the pull is concerned, into two or more units or single-joint muscle (Fig. 15.1). The tendons of the flexor profundus digitorum, for example, pass through several pulleys formed by fibrous sheaths. The direction of the pull is different for each joint and varies for each joint according to the position of the bones. The direction in each case is however determined, by a straight line between the centres of the pulleys on either side of the joint (Fig. 15.2). The direction of the pull in any of the segments would not be altered by a change in the position or origin of the muscle belly above the proximal pulley.

The joint may be modelled as a kinematic pair (a connection of 2 neighbouring segments), with given degrees of freedom (DOF) or independent variable defining the position (e.g., a ball and socket joint has 3 DOF, while a hinge joint has 1).

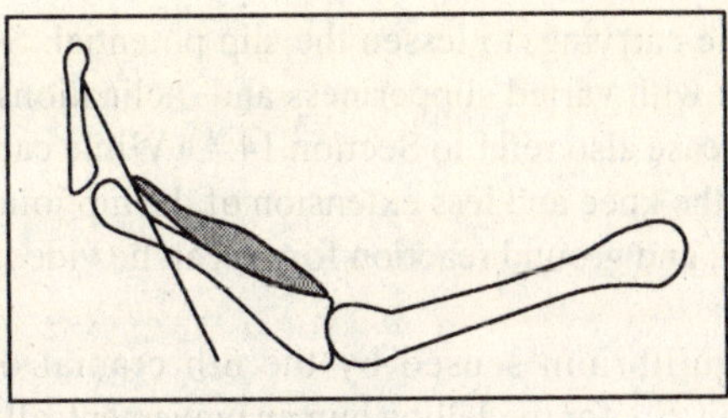

Fig.15.1

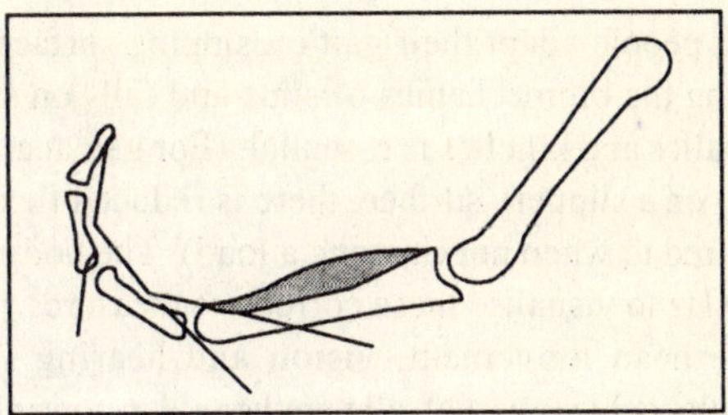

Fig.15.2

Action of the muscle pull on the tendon

In those regions where the muscle fibres are parallel or nearly parallel to the direction of the tendon, the entire strength of the muscle contraction acts in the direction of the tendon.

The work of the two muscles depends on the size of the contraction and on the tension exerted. It does not depend on the lever arm in very small contractions or in a summation of such contractions and therefore even for large contractions. In the first muscle a large load is moved through a short distance and in the second muscle a lighter load is moved through a greater distance.

Levers

The majority of the muscles of the body act on bones using the same principle as levers. Levers of the III class are the most common, for example, the action of the biceps, and the brachialis muscles on the forearm bones. Levers of the I class are found in movements of the head where the occipito-atlantal joint acts as the fulcrum and the muscles on the back of the neck as the power. Another common example is the foot when one raises the body by contracting the gastrocnemius and soleus. Here the ankle joint acts as the fulcrum and the pressure of the toes on the ground as the weight. This is frequently, though wrongly, considered a lever of the II class. If one were to stand on one's head with the legs up and with a weight on the plantar surface of the toes, it is easy to see that we would have a lever of the I class, if the weight were raised by contraction of the gastrocnemius muscle. If the fulcrum rests on the earth, the power must either directly or indirectly push from the earth or be attached to the earth, either by gravity or otherwise if it pulls toward the earth. If the power were attached to the weight, no lever action can be obtained. There are no levers of the II class represented in the body.

Muscle models may be cross-bridge models (sarcomere-microscopic, conventional cross-bridge and unconventional cross-bridge), fibre models and whole muscle-macroscopic models (viscoelastic and black box). *Dynamic analysis* is the analytic simulation of movements of the system considering the forces, torques and kinematics of the system (Section 14.4). *Forward dynamics* uses the geometry and kinematics of the system as input, and purveys forces and torques as outputs. *Inverse dynamics* starts from the forces and torques and determines the kinematics and geometry of the system.

RCOF

The ratio of shear to normal foot forces as the instantaneous 'friction used' may be denoted on 'dry surfaces' as the required *coefficient of friction (RCOF)*. Carrying a load affects the RCOF when descending a ramp (e.g., of inclination 10°) but not on level walking, more so in women. In other words, people adapt their gait on sloping surfaces, while carrying, to lessen the slip potential. So, for studying the biomechanics of slips and falls on surfaces with varied slipperiness and inclinations, gait kinematics and kinetics is essential. (For gait analysis, please also refer to Section 14.4) While carrying a load, on a slippery surface, there is reduced flexion of the knee and less extension of the hip joints (as compared to when not carrying a load). The body motion and ground reaction forces can be videotaped at 350 Hz to visualise these corrective measures.

The head movement, vision and hearing (and equilibrium sensed by the 8th cranial or the vestibulocochlear nerve), all together aid in controlling gait. So, for modelling human movement, all these factors have to be taken into consideration.

Johansson and Magnusson (1991) have argued that the study of posture dynamics is important not only to understand disorders of impaired equilibrium and protective reactions to unexpected displacements of the human body, but also to the design of prosthesis and functional neuromuscular stimulation as aids to patients with impaired postural stability and locomotion. Coordinated control of different body segments is a complex aspect of motor behaviour, and is made possible by the multiple degrees of freedom of the controlled system. Several interacting subsystems are involved in the dynamics of human posture and locomotion, including the skeletal, neuromuscular, and sensory systems. Human posture control is maintained by somatosensory, vestibular, and visual feedback, integrated within the locomotor and central nervous systems. Studies of posture dynamics and stability therefore entail study of the mechanical aspects of the human body, its sensory systems, and the principles governing coordination in motion control. Johansson and Magnusson, in their paper have reviewed some of the research done in the field of human posture dynamics, including such topics as biomechanics, equilibrium, stability, motion coordination, neural feedback, neural control systems modelling, motion strategies, optimality of motion, and adaptation. We will now consider experimental approaches and theoretical models, as well as the gap between them.

15.4 Biological Control Systems

Control of human movements, as mentioned before, is very complex. Common control systems are *open loop* (which uses a prestored trajectory and the model of the system to control the system) or *closed loop* or *error driven* (which uses information about the output to correct the control parameters to minimise errors between the desired and actual trajectory). There are also *reference based open loop control systems*, where the control system precomputes and stores control signals and executes the desired motor task in real-time.

Petrofsky and Philips (1985) have critically reviewed closed-loop (feedback) control of the skeletal muscle. They have examined the advantages and disadvantages of the open-loop control system as compared to the closed-loop control system in general, defined the problem, and outlined their approach. They defined the muscle structure and function at the level of the motor nerve, neuromuscular junction, and sarcomere. Time delays, power and efficiency, fatigue, and other effects were also discussed in relation to the development of closed-loop control. They then proceeded to review biological sensors and finally integrated this information by reviewing the body's own closed-loop control system. They also critically reviewed various approaches to the mathematical modelling of muscle. The control problem (in general) was reviewed with particular emphasis on contemporary control systems engineering.

Essential to closed-loop control of paralysed skeletal muscle is sensor technology. Therefore, they also reviewed external mechanical sensors. Specifically, potentiometers and Hall effect sensors, capacitive force transducers, inductive displacement transducers (LVDTs) (please see Section 1.5), and various position resolvers are discussed. Finally, they reviewed the application of closed-loop control of skeletal muscle to the human being. The focus of this section was the paralysed individual: past progress and future directions.

Agarwal and Gottlieb (1984 a and b) believe that the engineering analysis and modelling of the neuromuscular system is an exciting and challenging area of research where we may learn something about the principles of controlling large and complex systems. They describe the most peripheral subsystem: the postural control loop. In the first part, they reviewed the input-output properties of the motoneuron pool, modelling of the motor unit action potential, and analysis of the interference electromyographic activity. Also included were the engineering techniques to quantitatively study electromyographic activity and its relationship to muscle force. The second part of this review focused on the biomechanical models of the muscle based on Hill's work as well as the modern sliding filament theory. This section also included models of the proprioceptive receptors: the muscle spindle organ and the Golgi tendon organ. In the final paper they discussed various models and simulations of the peripheral loop and their applications to study physiological system behaviour such as tremor.

15.4.1 Analytic Controllers

Analytically, controllers can be simulated with mechanical and mathematical tools. Locomotion is often decomposed into a component in the sagittal plane, and smaller components in the frontal and horizontal planes. Using a smaller number of links is necessary for simulating gait. The human leg is represented by a double, planar pendulum with a moving hanging point. The rest of the body is denoted as a single force acting towards the moving hanging point (hip) called the hip joint torque. Two pairs of muscles acting around the hip and knee joints actuate the leg. Both monoarticular and biarticular muscles add to the joint torque.

15.4.2 Non-analytic Controllers

Although the analytic model of the body and extremities has its uses, it is complicated, the trajectories are not known in advance, and the driving resources are user-specific and time-variant. Therefore, real-time control using analytical tools is practically impossible.

As mentioned before, in non-analytic controllers, the operation is modelled on a set of mappings between the inputs and outputs.

Here the set of production rules (If– then, kindly see Section 19.4.2) is produced by the various neural networks like the fuzzy variant of MLP (Section 6.4.2) or the radial basis function (RBF).

Naturally, a combination, fusion, sandwich or hybrid technique is often preferred. A common example is the *Feedback error learning (FEL)* technique where mapping is used to replace the estimation of parameters within the feedback loop in a closed-loop control scheme. FEL is a feed-forward ANN (Section 19.4.2) structure that can be trained to learn the inverse dynamics of the control object.

15.5 Further Reading

[1] Bootsma-van der Wiel, A., Gussekloo, J., De Craen, A.J., Van Exel, E., Bloem, B.R., and Westendorp, R.G. Common chronic diseases and general impairments as determinants of walking disability in the oldest-old population. *J Am Geriatr Soc*. 50:1405–10. 2002.

[2] Tomović R., Popović , D. and Stein, R. B., eds. *Non-analytical Methods for Motor Control.* Singapore: World Sci. Pub. 1995.

[3] Latash, M. L. *Control of Human Movement.* Champaign, IL: Human Kinetics Publishers. 1993.

[4] Johansson, R., Magnusson, M. Human postural dynamics. *Crit Rev Biomed Eng* 18(6):413–37. 1991

[5] Petrofsky, J.S., Phillips, C.A. Closed-loop control of movement of skeletal muscle. *Crit Rev Biomed Eng* 13(1):35–96. 1985.

[6] Agarwal, G.C., Gottlieb, G.L. Mathematical modeling and simulation of the postural control loop. Part III. *Crit Rev Biomed Eng* 12(1):49–93. 1984.

[7] Agarwal, G.C., Gottlieb, G.L., Mathematical modeling and simulation of the postural control loop. Part II. *Crit Rev Biomed Eng* 11(2):113-54. 1984.

[8] Agarwal, G.C., Goodarzi, S.M., O'Neill, W.D., Gottlieb, G.L. Time series modeling of neuromuscular system. *Biol Cybern* 51(2):103–11. 1984.

[9] von Foerster, H. ed. *Cybernetics of Cybernetics*. Sponsored by a grant from the Point Foundation to the Biological Computer Laboratory, University of Illinois, Urbana, Illinois, 1974. [Now again available in re-published form]

10] Wiener, N. *Cybernetics, or Control and Communication in the Animal and the Machine.* Cambridge, Massachusetts: The Technology Press; New York: John Wiley & Sons, Inc., 1948.

15.6 Questions

1. Descriptive Questions

1. Why is modelling necessary for manufacturing an implant ?
2. What are the various types of models available for limbs ? Discuss their relative merits and demerits.

2. Write Short Notes on

1. Cybernetics
2. Modelling of implant
3. Limb modelling
4. Analytic controller
5. Non-analytic controller

3. Multiple Choice Questions:

Choose the single best response.

1. Models of muscles may be all except
   (A) sarcomere-microscopic
   (B) sarcomere-macroscopic
   (C) conventional cross-bridge
   (D) unconventional cross-bridge

2. A free rigid body has DOF
   (A) 1

(B) 2
(C) 3
(D) 6

3. In hierarchical control there is
(A) multilevel control
(B) vertical decomposition of the system
(C) both
(D) none

4. In kinematic analysis, analytic simulation of the movements of a system is done considering
(A) positions
(B) velocities
(C) accelerations
(D) all

5. The time histories of the instantaneous configuration of the system, in response to the control signals is known as
(A) trajectory
(B) plant
(C) state
(D) body

Answers

1. B; 2. D; 3. C; 4. D; 5. A

16

Principles of Feedback Control and Fluid (Blood/Air) Mechanics in Biology

16.1 Introduction

In the human physique there are thousands of control systems essential for homeostasis. Genetic systems operate in each cell to control both intracellular and extracellular functions. Regulation of O_2 and CO_2 concentrations in ECF is an excellent example of numerous control systems working in synchrony. Here, the respiratory and nervous systems work in harmony. Most control systems (cardiovascular, humoral or endocrine and neural) of the body operate by negative feedback.

16.2 Homeostasis

Walter B Cannon had coined the term *homeostasis* to express 'the various physiologic arrangements which serve to restore the normal state or *milieu interior*, once it has been disturbed'. A common example is the maintenance of the acid–base balance by the buffering electrolytes in our body fluids, mediated through the lungs (short term) and kidneys (long term). The whole gamut of neuro-endocrine functions act through such mechanisms (mostly short or long loop negative feedback, and occasionally some feed forward principles).

16.2.1 Feedback Control

The gain of a negative feedback is calculated as Gain = Correction /Error. For instance, the thermostat control of man has a feedback gain of –33. In other words, the thermostat of our body can correct to 33 times (in the opposite direction), a change in environmental temperature, compared to body temperature. For instance, if the room temperature changes from 20 °C to 53 °C (an alteration of 33 °C), the body can maintain the same temperature, at the most changing from 37.1 °C to 38.1 °C (a 1 °C difference).

16.2.2 Feedforward Control

Feedforward (positive or vicious circle) control systems anticipate changes. Some body movements occur so rapidly that nerve signals do not have sufficient time to travel from the periphery to the brain and back to control them. So, the brain uses feedforward control to cause the required muscle contractions. Sensory nerve signals from the moving parts inform the brain retrospectively whether the expected movement has occurred or not. In case it has not occurred, the brain corrects the feedforward signals that it sends to the muscle next time. This is known as adaptive control or delayed negative feedback.

16.3 Fluid Dynamics

Fluid dynamics (mechanics) in our body is mainly the study of the flow of blood (hemodynamics) and air (in the lungs). While the circulatory system (blood) is a positive pressure pump, the respiratory system (air) is a negative pressure pump.

Interestingly, physiological fluid dynamics deals with the flow of fluids inside living organisms, such as in the respiratory or cardiovascular systems, while zoological fluid dynamics deals with the flow of fluids external to living bodies, such as while swimming, flying or bioconvection.

16.3.1 Hemodynamics

Normally blood flow is laminar (streamline). However, above a certain critical velocity, the flow becomes turbulent and produce sounds. The probability of turbulence is denoted by the *Reynold's number* $\mathrm{Re} = \rho D \dot{V} / \eta$, where ρ (poises) is the density of blood, D (cm) is the diameter of the vessel, $\dot{V}$ is the velocity of the flow, while η is the viscosity of blood. The higher the Re, the more is the chance of turbulence. Usually, if Re is below 2000, the blood flow is usually not turbulent, and above 3000, it is almost invariably turbulent.

Low Reynolds number fluid dynamics

One of the areas of biology in which applied mathematics has made a number of important contributions is the field which studies the swimming of microorganisms. In the world of Stokes flow relevant to motile bacteria, where motion is primarily geometrical rather than dynamical in origin, self-propulsion by rotating helical flagella or waving elastic flagella is a central subject. Another area of fluid mechanics, related to microorganisms and their resisting white blood cells is the *chemotactic* response of individual cells to external chemical gradients. This is a phenomenon that involves not only complex biochemical pathways but also the physics of diffusion and propulsion. As some classic experimental work on *E. coli,* has revealed, regulation of the flagellar motors leads to a swimming motion that involves 'runs' and 'tumbles'. As a consequence, cells execute a random (or biased random) walk.

Blood flow through a vessel is determined by the pressure gradient and vascular resistance as $Q = \Delta P/R$, where Q is the blood flow, ΔP is the pressure difference between the two ends of the vessel and R is the vascular resistance. In the pulmonary circulation, ΔP is much lower than that in the systemic circulation, while blood flow (BF, Section 16.5) is the same as that in systemic circulation, therefore, total pulmonary vascular resistance is very much lower than the systemic vascular resistance. *Poiseuille's law* or *Poiseuille Hagen formula* describes the relation between BF, ΔP, η, radius of the vessel (r) and length of the vessel (L).

$$BF = \Delta P \times \left(\frac{\pi}{8}\right) \times \left(\frac{1}{\eta}\right) \times \left(\frac{r^4}{L}\right)$$

It states that vascular *resistance* (α) is directly proportional to the viscosity of the blood and the length of the blood vessel and inversely proportional to the radius of the blood vessel, raised to the power of four. So, if the radius of a blood vessel is increased to twice its original width, resistance will decrease to 1/16 of the original and BF will go up by 16 times, if ΔP remains unchanged. In other words, smaller vessels have greatest resistance while large vessels have little resistance to BF. *Conductance* is the reciprocal of resistance and is a measure of ease of BF.

Increased resistance due to high viscosity may be due to raised packed cell volume (PCV) or hematocrit, as in polycythemia. Contrarily, anemia will lead to hyperkinetic circulation. Vascular *compliance* or *capacitance* is the total quantity of blood that can be stored in a given part of the circulation system, and is measured by (↑in volume / ↑in pressure). The greater the compliance, the more easily can the system be distended by pressure. So, compliance can also be measured as distensibility multiplied by volume.

In the brain, cerebral blood flow (CBF) is measured by the invasive *Kety method*, using N_2O, following the *Fick principle*. Non-invasively, it may be measured by the *tetrapolar impedance method.* In the Kety method, CBF = $Q_x / ([A_x] - [V_x])$, where x is N_2O, and, Q denotes its perfusion, while A and V denote its arterial and venous concentrations respectively. In the case of kidneys, x is para amino hippuric acid (PAH). By the non-invasive method, BF may be denoted as $\Delta V = \rho L^2 \Delta Z / Z_0^2$, where ρ is the resistivity of the blood (Ω cm), L is the distance between the measuring electrodes, Z_0 is the basal impedance (Ω) and ΔZ is the change in impedance. The other instruments for measuring blood flow include *electromagnetic flowmeters and Doppler flowmeters* (Section 17.3.3). In the case of an electromagnetic flowmeter, the voltage generated in the conductor (blood) moving through a magnetic field, is proportional to the speed of the BF. As the voltage is proportional to the volume flow, it can be measured to give the value of the flow. A sealed watertight chamber (plethysmograph) is also used for measuring the BF.

Unlike in rigid tubes and homogeneous fluids, the relation between pressure and flow in blood vessels is not linear. The pressure at which blood flow stops is known as the *critical closing pressure.*

Laplace's law states that the distending (transmural) pressure (P in dynes/cm²) in a distensible hollow object, at equilibrium, is equal to the tension (T in dynes/cm) in the wall, divided by the two principal radii of curvature (r_1 and r_2 in cm). In a sphere, r_1 and r_2 are equal (r), and, in a cylindrical blood vessel, one radius is infinite, therefore:

$$P = T\,(1/r_1 + 1/r_2) = 2T/r = T/r$$

Therefore, less wall tension is necessary to balance the distending pressure in a smaller vessel. While the tension (with normal BP) in the human aorta is 170,000 dynes/cm, in a capillary it is about 16 dynes/cm.

Blood gas analysis

Blood gas analysers commonly measure pH (reference range: 7.35–7.45; i.e., a blood pH of 7.3 is actually considered to be a state of acidosis in our body, while in chemistry, it will be taken as alkaline pH); pCO_2 (35–45 mm of Hg or 4.65–6 kPa) and pO_2 (arterial blood: 80–100 mm Hg or 10.6–13.3 kPa; venous blood: 25–40 mm Hg or 3.3–5.3 kPa). However, to monitor the acid–base balance, blood HCO_3^- (bicarbonate) and BE (base excess) are also measured. Albeit pH may be measured from venous blood, pO_2 in the arterial blood is more important (that will decide the supply of oxygen to the tissues).

The blood pH is commonly measured with a glass electrode (the other reference electrode is usually of calomel, connected by a KCl bridge) and is derived from the Nernst equation (Section 1.2.1). Usually the change of one unit of pH denotes a potential change of 58.2 mV at 20°C and 62.2 mV at 40 °C. For long-term monitoring, microcapillary glass electrodes are used. The buffer solutions used for measuring blood pH are 0.025 molar $KH_2(PO_4)$ having a pH of 6.84 at 38°C and 6.881 at 20°C; and 0.01 (M) $KH_2(PO_4)$ with 0.04(M) Na_2HPO_4, having a pH of 7.416 at 38 °C and 7.429 at 20°C.

The pCO_2 in blood is measured anaerobically as:

$$pCO_2 = \text{barometric pressure} - \text{water vapour pressure} \times \% \, (CO_2 / 100)$$

At 37 °C the water vapour pressure at sea level (760 mm Hg atmospheric pressure) is 47 mm Hg, so, 5.7% pCO_2 corresponds to pCO_2 of 40 mm Hg.

For understanding the acid–base equilibrium, we must know about the Henderson Hasselbalch equation and the dissociation constant (*K*). *K* is expressed by $[H^+][A^-]/[HA]$, where A^- is the anion or base.

$$pH = pK + \log\frac{[\text{base}]}{[\text{acid}]} \quad \text{or} \quad \log\frac{[\text{salt}]}{[\text{acid}]}.$$

The [] denotes the concentration. Usually the HCO_3^- (22–26 mM or mmol/L): H_2CO_3 (around 1.2 mM) ratio is 20:1. At 37 °C, pK for plasma is 6.1, therefore, $[HCO_3^-]$ is calculated by determining pH and pCO_2 by the formula:

$$pH = pK\,(6.1) + \frac{\log[HCO_3^-]}{0.03\,pCO_2}$$

Total CO_2 (mM) is calculated by the formula

$$([HCO_3^-] + 0.03 \times pCO_2)$$

Base excess (BE) is:

$$BE = (1 - 0.0143 \times Hb)[HCO_3^-] - (9.5 + 1.63\,Hb) \times (7.4 - pH) - 24$$

where Hb is the concentration of haemoglobin.

Commonly Clark type oxygen electrode measures blood pO_2, where oxygen from the blood diffuses across the membrane into the electrolyte solution and is reduced at the platinum cathode (reaction: $O_2 + 2H_2O + 4e^- \rightarrow 4OH^-$ and simultaneously, the silver anode reaction: $4Ag \rightarrow 4Ag^+ + 4e^-$). Oxygen is usually reduced at the polarising voltage of 0.6 to 0.9V.

Oxymetry

In oxymetry, the percentage saturation of oxygen in the circulating blood is measured as:

$$\frac{[HbO_2]}{[HbO_2]+[Hb]}$$

where [Hb] is the concentration of deoxygenated haemoglobin and $[HbO_2]$ is the concentration of oxygenated haemoglobin. This may be measured *in vitro* (transmission or reflection oxymetry) or *in vivo*, but the latter is more common. Here the pinna or the external ear, or the finger tip is used for measurements. Another (more accurate way is to directly measure through cardiac catheterisation). In the case of ear measurements, the transmission of light through the pinna is used to measure the oxygen saturation. In the case of fingertips, spectrophotometric analysis of the transmitted light through the fingertip measures the percentage saturation (*S*) of oxygen by the formula:

$$S = A - B\left(\frac{\alpha_{650}}{\alpha_{805}}\right)$$

where α denotes the absorption coefficient of the whole blood at that wavelength (nm) of light and *A* and *B* are constants related to absorption coefficients of respectively Hb and HbO_2.

16.3.2 Dynamics of air in the lungs

Movement of air in and out of lungs is caused by changes in pleural pressure (pressure of the fluid in the narrow space between visceral pleura and parietal pleura, ranging from –5 cm of water at the beginning of inspiration to –7.5 cm H_2O at full inspiration) and alveolar pressure (air pressure inside the lung alveoli), which varies from –1 cm H_2O during inspiration to +1 cm H_2O during expiration and being zero when the glottis is open and there is no movement of air. Lung *compliance* is the change in lung volume, for each unit change in transpulmonary pressure ($\Delta V/\Delta P$). Water molecules have a strong affinity for each other, so the water surface lining the alveoli causes surface tension elastic (contractile) force of the entire lungs. *Surfactant* reduces the work of breathing (raising compliance) by decreasing alveolar surface tension (ST). Smaller alveoli have a greater tendency to collapse. Collapse pressure = (2 × ST) / radius. So, when alveoli have half the original radius, collapse pressures are doubled. Surfactant 'interdependence' and lung fibrous tissue are important for 'stabilising' sizes of alveoli.

The ratio of the pulmonary ventilation and blood flow (perfusion), called the *ventilation perfusion ratio* of the whole lung is about 0.8, but varies according to the region of the lungs.

The amount of air that moves into the lungs with inspiration (or moves out with expiration) is called the *tidal volume* (V_T). The air inspired with maximal inspiratory effort above the V_T is called the *inspiratory reserve volume* (IRV). The volume expelled by active expiratory effort after V_T is the *expiratory reserve volume* (ERV), and the air left in the lungs after maximal expiratory effort is called the *residual volume*.

The largest amount of air that can be exhaled after a maximal inspiratory effort is called the *vital capacity (VC)* and is given by the sums of IRV, V_T and ERV. The fraction of VC expired during the first second (FEV_1), also known as timed vital capacity or FEV_1 (taken each second for the first 3 seconds, respectively 80–85%, 93–94% and about 97%) may be reduced despite normal VC in obstructive airway diseases. The zone where no exchange of gas takes place is called the *respiratory dead space* or V_D (anatomical or actually, areas where exchange never occurs and physiological or functional, moments when the body is in resting or inactive conditions). Bohr's equation states that:

$$P_{E_{CO_2}} + V_T = P_{a_{CO_2}} \times (V_r - V_D) + P_{I_{CO_2}} \times V_D$$

where P stands for partial pressure of the gas (CO_2) in expired (E) or inspired (I) air or arterial blood (a). The last term (product of the inspired gas partial pressure P and V_D) is so small that it can be safely ignored. In that case, for instance, with expired and arterial pCO_2 being respectively 28 mm Hg and 40 mm Hg, and V_T being 500 mL, V_D will be 150 mL.

The diffusion capacity of the lungs for a given gas is directly proportional to the size of the alveolocapillary membrane and inversely proportional to its thickness. The diffusion capacity (D_L) of carbon monoxide (CO) is measured as:

$$D_{L_{CO}} = \frac{\dot{V}_{CO}}{P_{A_{CO}}}$$

where the numerator denotes the amount of CO entering the blood and the denominator reflects the partial pressure of CO in the alveoli (however, in smokers, it has to be subtracted by the substantial value of pCO in pulmonary capillaries, unlike the near zero value in non-smokers).

Pulmonary function test (PFT)/Spirometry

Usually ventilation (spirometry), distribution (indirect measurement with nitrogen washout) and diffusion (diminution of a small amount of CO mixed with inhaled air) measurements are done.

The classical water sealed spirometer can adequately carry out measurements for vital capacity and tidal volume. It comprises an upright water-filled cylinder with an inverted counter weighted bell. Breathing into the bell alters the volume of gases trapped inside, and is converted to vertical motion, which, in turn is recorded on the moving drum of a kymograph.

The other variants are wedge spirometer, ultrasonic spirometer, pneumotachometer (measures the instantaneous rate of volume flow of respired gases, may be Fleisch, Venturi, or turbine types).

Smaller peak flow meters can be used for bedside measurements.

In digital varieties, the flow-volume curves are also calculated and printed.

A complete digitised pulmonary function analyser consists of a nitrogen analyser, a vacuum pump, an *xy*-recorder, a pneumotachometer, a digital monitor and an electronic circuitry. An impedance pneumograph can purvey continuous volumetric recording of respiration.

Respiratory gas analysis

Commonly the gases measured clinically are CO, CO_2, N_2O and halothane in mixtures of respiratory or anesthetic gas samples. Infrared gas analysers utilise the property of some gases to absorb specific infrared wavelengths. It can be used for all the gases mentioned above.

Paramagnetic oxygen analysers and polarographic oxygen analysers are often used for measuring the oxygen level in the samples.

Thermal conductivity gas analysers too may be used for most gases.

An ionisation technique based nitrogen analyser is also used.

Halothane vapour strongly absorbs UV light at 200 nm and therefore halothane meters are based on UV absorption principles.

Appendix

Some of the recent areas of research in fluid dynamics (e.g., by the Biological Fluids Group, DAMTP, University of Cambridge: www.damtp.cam.ac.uk) are:

1. Blood flow and wall shear stress in arteries (atherosclerosis)
   Coronary arteries (tubes of time-dependant curvature)
   Bypass grafts
   Dynamics of endothelial cells
2. Flow and oscillations in collapsible tubes (blood vessels, airways and urethra)
   1-D, 2-D and 3-D models: coupled solutions of Navier-Stokes and shell equations.
   Laboratory experiments
   Giraffe jugular vein
3. Pulmonary fluid dynamics
   Gas flow and mixing in the airways at high frequency
   Surface tension and surfactant effects
4. Coupling of hydrodynamics and muscle mechanics
   Peristaltic pumping in the ureter
5. Hydrodynamic phenomena in populations of swimming microorganisms
   Bioconvection (algae and bacteria)
   Chemotaxis in flow fields

16.4 Further Reading

[1] Ganong, W.F. *Review of Medical Physiology*, 20th ed. Prentice Hall Intl (Inc.). 2001.
[2] Guyton, A. C. and Hall, J. E. *Textbook of Medical Physiology*, 10th ed. W B Saunders. 2001.
[3] Raja Rao, C. and Guha, S. K. *Principles of Medical Electronics and Biomedical Instrumentation.* Universities Press (India) Ltd. 2001.
[4] Khandpur, R. S. *A Handbook of Biomedical Instrumentation.* Tata McGraw Hill. 2002.
[5] Webster, J. G., ed. *Encyclopedia of Medical Devices & Instrumentation.* NY: John Wiley & Sons.1988.

16.5 Questions

1. Descriptive Questions

1. What sorts of feedback are necessary for maintaining homeostasis in our body?

2. What is ventilation perfusion ratio? What are the parameters for measuring ventilation? What are the common instruments used for this purpose?

2. Write Short Notes on

1. Dead space measurement
2. Negative feedback
3. Blood gas analyser and oxymeter
4. Pulmonary compliance
5. FEV_t

3. Multiple Choice Questions

Choose the single best response.

1. Specific gravity of whole blood is
   (A) 1.025–1.035
   (B) 1.035–1.045
   (C) 1.045–1.065
   (D) 1.065–1.075

2. Homeostasis may be maintained by
   (A) negative feedback
   (B) positive feedback
   (C) feed forward
   (D) any of these principles

3. Haemodynamic studies are important for all except
   (A) understanding flow and pressure changes of blood
   (B) blood gas analysis
   (C) neutrophilic chemotactic studies
   (D) bruits and murmurs

4. The following is not correct about surfactant:
   (A) It helps in keeping alveolar surface tension low
   (B) It causes atelectasis
   (C) It is not well secreted in premature neonates
   (D) Its absence may cause hyaline membrane disease.
5. Timed vital capacity (FEV_1) may be reduced despite normal vital capacity
   (A) in asthma
   (B) due to reduced airway resistance
   (C) in bronchodilation
   (D) any of the above

Answers

1. C; 2. D; 3. B; 4. B; 5. A

Section 6

Imagology

In this section, we will discuss the various imaging modalities available for medical diagnosis and treatment.

In Chapter 17, ultrasonography and colour Doppler technologies are discussed.

Chapter 18 deals with X-rays, CT, PET, SPECT, MRI, fMRI scans and NMR spectroscopy.

17

Ultrasonography, Colour Doppler

17.1 Introduction

Ultrasonography (USG) is an excellent non-invasive (albeit operator dependent) diagnostic modality, which can distinguish interfaces between soft tissues. Piezoelectric crystals are used to generate and detect ultrasound (US) waves (frequency > 20 kHz, i.e., ultra or above human audible range). These crystals can develop electric potentials on definite crystal surfaces, when exposed to mechanical strain. Materials with high mechanical Q factor are suitable as transmitters, while those with low Q value and high sensitivity are preferred as receivers and for non-resonance applications. Synthetic lead zirconate titanate (PZT) and polyvinyldience fluoride (PVDF) crystals are far better than natural quartz crystals for a frequency of up to 15 MHz, or now, even up to 50 MHz. At higher frequencies, quartz is usually preferred since it has better mechanical properties. The surfaces of synthetic crystals are silvered for making external electric connections. The three important parameters for optimising transducers for numerous applications are: frequency, active element diameter and ability to focus.

In ultrasonography, sound waves of high frequency are used to produce images. When high frequency ultrasonic waves pass through different parts of our body, these parts respond differently. Many tissues reflect these sound waves partially and transmit the rest, which are then reflected back from more deep structures. The reflected waves are measured and depending on the time it takes for them to return, the depth of the echo is decided – the intensity decides the grayness of the area. Ultrasound is used to examine many parts of the body, specifically to observe the fetus, check for other gynecological abnormalities and to examine the abdomen, orbits, thyroid glands, breast, and testes. This examination is not very useful for evaluation of diseases of the brain and chest. Ultrasonography of the abdomen requires prior preparation which includes bowel evacuation and an empty stomach.

Table 17.1 Characteristic impedance for some of the physiological materials

Material	Impedance (× 10^5 g/cm^{2}s = × 10^6 kg/m^{2}s = Rayls ′ 10^{-6})
Water	1.48
Fat	1.38
Muscle	1.70
Kidney	1.62
Air	0.0004

For an US beam impinging on the interface perpendicular to the surface, the *reflection coefficient (R)* is given by $((Z_2-Z_1)/(Z_2+Z_1))^2$, where Z_1 is the characteristic impedance of of the material in the beam path before interface and Z_2 is that along the beam path after interface. If the value is not squared, we have the ratio of amplitude of reflected wave to that of incident wave, and this amplitude denotes the impedance change at the interface. This impedance change is represented as gray scale values for image formation. Some of the representative values of impedance are tabulated in Table 17.1. For calculating the *depth* (*d*), we use the formula

$$d = \frac{\Delta t}{2}.v$$

where ΔT is the time delay between the input signal and the output (returned signal) and v is the ultrasonic propagation velocity.

New advances in ultrasound technology has allowed higher resolution scanning for small areas such as the superficial ligaments and tendons of the body, the use of Doppler waves which allow examination of blood vessels and the use of 3-D ultrasound which allows three-dimensional studies of various parts of the body.

Nearly all techniques that are used today in the field of medical ultrasound diagnostics function according to the principle of the so-called *echo-pulse method*. To transmit an acoustic wave, an electric field is applied to the piezoelectric crystals of the transducer which will afterwards generate a pressure wave. The *pulse* duration is about 1 μs. After transmission, the transducer is switched to the receive mode. The transmitted sound wave propagates through the body and is reflected at the internal boundaries between tissues with different acoustic impedance. The part of the beam that comes back to the piezoelectric crystal of the transducer is referred to as *echo* and generates a new electric signal at the crystal. The time between the pulse transmission and the receiving of the separate returned echoes is measured. The distance covered by the sound wave z is given by the product of the sound velocity c and the delay time t that lies between transmission and receipt of the signal. By dividing this product by 2, the precise location of the examined organ in relation to the position of the transducer can be determined.

$$z = \frac{ct}{2}$$

A delay time of 0.13 μs means that the structure from where the echoes are reflected is situated at a depth of 10 cm. Modern ultrasound systems are not only able to transmit between 3,000 and 5,000 sound waves per second but are also able to receive, calculate and transform the data of the returned echoes into images within the same period of time.

Echoes returning from deeper structures normally have lower energy (amplitude) than echoes returning from superficial structures and therefore need more amplification to compensate for attenuation of the signal. This is why boundaries which have the same reflection coefficient but are located in different depths return echoes with different amplitudes. Uniform echo amplitude throughout the image is achieved by the appropriate amplification of all signals that are returned to the crystal of the transducer. The time gain control (TGC) or depth gain control (DGC) can be set by the user himself according to specific needs.

The oldest and most simple imaging mode is the A-mode. In this imaging mode, the amplitudes of the pressure waves generated by the transducer are displayed on a cathode-ray oscillograph with regard to the position of the boundaries in the examined tissue. It conveys one-dimensional information, and is still applied in ophthalmology (determination of the thickness of the cornea) and in otolaryngology (non-invasive diagnostics of the paranasal sinuses).

In contrast to the A-mode, where the amplitudes are represented in the form of peaks, the amplitudes in B-mode are represented as pixels. The *brightness* of the pixels corresponds to the intensity of the electrical signals, i.e., the strength of the echoes. Stronger echoes are represented by brighter pixels. This explains why the B-mode is also called *brightness-mode*. Modern diagnostic ultrasound systems normally offer 256 different brightness levels – the so-called grey scales (The human eye is able to distinguish up to 20 different grey levels on the same image). On the screen, the separate pixels are displayed in the form of a line. Before each transmit pulse, the beam is shifted accordingly. By arranging the new lines according to the position of the structures from where the echoes have been reflected, a two-dimensional image can be achieved.

In M-mode, the so-called *motion-mode*, the beam is not shifted as in B-mode but is held in a fixed position over the tissue to be examined. Now, the single lines generated on the basis of the echoes returned are arranged on the time-axis, in order to show how the returned echo data change over time. With a penetration depth of 15 cm and a scan time of 0.2 μs, up to 5,000 images can be constructed per second. This also allows the documentation of very quick motions such as that of moving heart valves. With the known scale of the graph and the time-axis, the precise documentation of motion, velocity and acceleration is possible in M-mode. This is why so much importance is given to this imaging mode in echocardiographic applications.

17.2 Ultrasonography (USG)

17.2.1 Diagnostic USG

Diagnostic ultrasonography is a technique that allows non-invasive visualisation of tissues within the body. The ultrasound machine consists of a box (approximately one foot square) attached to a screen. A 3/4 inch diameter cord leads from the box to a specialised probe. The probe emits sound waves of specific lengths, which are not detectable by the human ear, in a beam the length of the probe and 1/25th of an inch wide. When the probe is placed on the skin, or within the rectum of an animal, the sound waves penetrate the tissues and are reflected back to the probe. The density of the tissues determines the number of waves that are reflected. The probe detects these waves and a computer within the ultrasound box converts the waves into a picture displayed on the ultrasound screen. The black and white picture depicts a cross-section of the tissues directly beneath the probe. Various probes are available, depending on the intended use. Probes used for most reproductive work in large animals are made so as to be easily held in the palm of the hand and inserted into the rectum.

For diagnostic use, the USG machine operates at less than 100 mW/cm^2. Diagnostic USG may be a continuous or pulsed wave. The continuous waves act through the Doppler effect and are used for fetal heart and blood flow monitoring. The incident US is scattered by red blood cells (rbcs) and the scattered wave is received by the second transducer. The frequency shift due to the moving scattering particles is proportional to the velocity of the particles. Alteration in frequency occurs first as the US arrives at the scattering particle, and second, as the US leaves it.

WHO study group on training in diagnostic ultrasonography

The WHO Study Group on Training in Diagnostic Ultrasonography (Essentials, Principles and Standards), met in Philadelphia, Pennsylvania, USA on March 22-26, 1996 to consider methods of improving the training and clinical performance of medical doctors and technologists using diagnostic ultrasound for medical applications.

The meeting was organised with the collaboration and financial support of the World Federation for Ultrasound in Medicine and Biology (WFUMB). The meeting was opened by Dr. A. Wasunna, Director, Programme on Health Technology WHO, Geneva, on behalf of the Director General of the WHO, Dr. Nakajima.

The study group represented a wide variety of expertise on the topic. In addition to the WHO Geneva Office representatives, the group consisted of physicians, sonographers, physicists and engineers from all over the world.

Several papers highlighting currently used training procedures and needs in different countries were presented by delegates from South East Asia, including China, Japan and the Indian subcontinent, Australia, North and South America, and Europe.

The ensuing discussion formed a background for a final document which was approved by the workshop participants. To prepare this document different working groups were formed; each group prepared its final draft for discussion. The drafts addressed the following topics:

- world situation: training in diagnostic ultrasonography
- specialty ultrasound including specific recommendations for individual specialties
- ultrasound training for sonographers and other allied health personnel
- curriculum for basic sciences and instrumentation
- training centres and resources needed
- recommendations

Recognising the importance of ultrasound as the fastest growing diagnostic imaging modality, the workshop participants emphasised the need for uniform training for sonographers and other allied health personnel such as nurses, midwives, radiographers, nuclear medicine technologists, physician's assistants and cardiac technologists. It has also been stressed, that as proper ultrasound diagnosis is very operator-dependent, appropriate training is essential.

One of the study groups developed a basic curriculum for physics and instrumentation. As this aspect of training appears to be of particular interest to medical and biological engineers, it is briefly summarised below.

The developed curriculum emphasises that understanding the basic principles of physics and knowledge of instrumentation are essential to achieve the maximum benefit of ultrasound technology for diagnosis and to avoid possible pitfalls that may render diagnosis useless. Accordingly, the proposed curriculum focused on two main objectives.

- To provide the ultrasound practitioner with sufficient knowledge of ultrasound fundamentals to effectively operate an ultrasound scanner, and to provide a diagnostically useful image with appropriate regard to issues of benefit and potential risks to the patients' safety.
- To provide an understanding of the nature of artifacts that may be present in the images formed in order to avoid misdiagnosis.

Basic physics, including bio-effects and safety, and imaging methods and equipment, including quality assurance, constitute the major topics of the curriculum. It was proposed that the basic curriculum be delivered within 20 hours of lectures. The advanced curriculum will require 20 additional hours of classes.

The developed curriculum stresses that basic physical principles relating to ultrasound and its interaction with tissues should be mastered. This knowledge must include basic terminology; physical relationships involving wave pressure amplitude, intensity, attenuation, perpendicular and oblique angle of incidence, scattering and the range equation. In addition, transducer construction and the effects of the sound waves produced should be discussed, together with methods of focusing. Issues associated with resolution and trade-off between resolution and tissue penetration must also be taken care of. Finally, artifacts, bio-effects, safety and the principles of scanning are to be included.

Level II training is to be tailored to the needs of individuals with an MD. Here the curriculum includes advanced physics of ultrasound imaging, terminology, bio-effects, safety, on-screen output display standards and their interpretation, ultrasound instrumentation, operation of the equipment, set-up and quality assurance. Instrumentation knowledge covers all B-mode imaging, pulsed wave and continuous wave Doppler, and colour and power Doppler.

The advanced level curriculum document further stresses the importance of understanding and appropriate knowledge of statistics and computer science. Finally, the document indicates the need for continuing medical education and follow-up training.

The document stresses that even established well-advanced teaching centres should be encouraged to invite external lecturers to ensure fostering of new ideas and to constantly improve the quality of the curriculum. In addition, the potential advantage of using distant learning and the multimedia approach was highly recommended.

Implementation of these curricula will require the active involvement of biomedical engineers and the development of appropriate curricula. Biomedical engineers are well- suited for this task as they usually combine excellent theoretical knowledge with hands-on practical experience.

The development of the curricula should take into account the fact that ultrasound is now considered to be the primary imaging modality for the detection of most gynecological, hepatic, biliary, pancreatic, spleenic and renal disease. Ultrasound diagnosis is also extensively used in the examination of the scrota, bladder and prostate. In addition, in many developing countries, sonography may have an important application as an epidemiological screening tool and as a diagnostic procedure in parasitic diseases such as amoebiasis, schistosomiasis and echinococcosis.

In light of the above, it appears that the delivery of the recommended curricula will generate a need for engineers who not only must possess the required professional background, but who are also able to develop exciting curricula, carefully tailored to a generally non-engineering audience.

Indeed, the workshop document stresses the need for highly qualified teachers. Teachers should regularly attend ultrasound meetings and conferences to ensure that they remain in touch with this rapidly developing field.

The proposed curricula also address the issue of a basic knowledge of biology. This knowledge is essential in understanding the origin of tissue-reflecting properties and for an appreciation of the issues involved in bio-effects and safety. Again, this would create a need for biological science instructors with appropriate physical acoustics backgrounds.

At present, it appears that there are a limited number of individuals who are adequately trained to teach the proposed curricula. This indicates a need to establish training support groups, comprising individuals who would combine scientific and didactic knowledge. Such individuals would also be able to provide appropriate pedagogical assistance which would enable engineers to deliver high-quality lectures tailored to the specific audience.

The development of intensive training programs for teachers is desirable and has already been provided in some countries, e.g., a prototype of such a program has been produced at the Jefferson Research and Education Institute, Philadelphia, Pennsylvania. In this program, individuals not only receive an update of the clinical and technical knowledge in the fields in which they are going to teach, but also instruction in the methods and techniques of teaching. These programs are a model for 'teach-a-teacher' programs and should be developed at major training centres throughout the world.

Regardless of their level of expertise, all teachers and practitioners of ultrasound must pursue continuing education through refresher courses, as well as updates on new technology. This is particularly important if individuals wish to expand their knowledge into new areas of interest.

In addition, (biomedical) engineers with a background in imaging instrumentation will be needed to provide appropriate technical back-up for the physicians, sonographers or other health allied personnel involved in ultrasound diagnosis.

A final version of the comprehensive report approved by the members of the Study Group is going to be published shortly and made available to all interested parties. The report concludes with a list of recommendations.

- Appropriate training of the users of diagnostic medical ultrasound equipment is the most important requisite for the improvement and rational use of diagnostic ultrasound in clinical practice. The use of ultrasound by individuals with inadequate competence adds to the likelihood of unnecessary patient examination and misdiagnosis, with substantial negative impact on the healthcare system.
- It is highly desirable that the purchase and use of medical ultrasound equipment is restricted to those who have acquired and passed an adequate training in diagnostic ultrasound. The Study Group reinforces the recommendation of the WHO Scientific Group on The Future Use of New Imaging Technologies in Developing Countries (TRS 723) that 'purchase of ultrasound equipment without provision for the proper training of an operator is contrary to a good healthcare practice and is unlikely to be cost-effective'.
- The Study Group strongly recommends that appropriate curricula, such as those included in their document, should be adopted for general, advanced and specialised training of medical doctors and allied personnel who use diagnostic ultrasound.
- The WHO, international, governmental, non-governmental organisations and professional organisations should be actively involved in the development of high quality training programs. This can be achieved by setting high training standards, organising and carrying out of training courses, certifying trainees, and organising continuing education programs for both teachers and program participants.

Probes used in diagnostic USG

One of the various probes used for diagnostic USG are the transducers (probes) which may be linear array, curved or convex array, sector probes or phased array. Apart from these electronic phased arrays, there may be mechanical sector probes (rotating, wobbling or annular).

In linear array probes, which are also known as parallel transducers, the single elements are arranged in the form of a straight line. The beams generated by this group run parallel to each other and are displayed on the screen in a rectangular form. With linear array probes, a high resolution can be achieved for nearly all depths. Linear array probes usually consist of between 60 to 196 single elements that are arranged in a distance of 1 to 4 λ and operate with frequencies in the range between 5 and 13 MHz. To focus perpendicular beams, linear array probes are often equipped with an additional acoustic lens. Curved linear array probes and convex linear array probes belong to the group of linear array probes and function according to the same principle. The only difference is that the elements are arranged in a curved line which leads to a fan-shaped ultrasonic field. As the line density decreases with greater distance from the transducer, the lateral resolution that can be achieved in deeper structures is quite poor. Curved arrays and convex arrays usually consist of at least 96 elements, have a radius of curvature of between 25 and 80 mm and work with frequencies in a range between 3 and 7 MHz and a scan angle of 60° to 90°. Sector probes have a radius of curvature of below 25 mm and therefore have a smaller surface area which leads to a smaller near field and a higher angle of distortion of more than 90°. They are, therefore, frequently used for the intercostal space-view in echocardiography and for transvaginal applications. The elements of phased array probes, like in the linear array probes, are arranged in a straight line. The difference is that the ultrasound waves are generated by all elements of the transducer and not by single groups. By timing the transmit pulses to the single elements, it is possible to generate a wave front that runs in a certain angle to the surface of the transducer and to create a sector-shaped ultrasonic field. A typical feature of phased array probes is their small surface area of 12 to 20 mm. Phased array probes normally consist of 64 to 128 elements, operate with a sector angle between 80° and 90° and use frequencies in the range between 2 and 7 MHz. Because of their highly complicated electronic nature, phased array probes are quite expensive and are therefore mainly used for cardiac applications or for transcranial ultrasound examinations. In contrast to the electronic phased array probes, ultrasound signals can easily be generated and processed by mechanical sector probes. Owing to their cost effectiveness, mechanical sector probes are more frequently used than phased array probes. Mechanical sector probes are either available with a rotating or with a wobbling transducer. These probes are equipped with a rotating device with 3 to 5 single crystals which are arranged in a certain angle to each other. The rotor is started by a small motor that is placed at the handle of the probe. The element in the probe window is activated, thus opening the view to a sector-shaped acoustic window. Afterwards, the next element moves into the probe window and produces the next image. In probes with wobbling transducers, a single crystal swings back and forth in a scan sector of 60° to 100°. In contrast to rotating transducers, wobbling transducers can easily be adjusted because they work with a single crystal and also allow the variable adjustment of the scan angle. Both methods remain restricted to a fixed focus zone because they only work with one or a few crystals. Annular phased array probes work with a wobbling transducer and contain typical features of both mechanical and electronic probes. Instead of a single crystal, these probes consist of several elements that are arranged in a concentric circle. The individual timing of the circles allows variable dynamic 2D-focussing.

The transvaginal pulse-echo probe comprises a single element pulse-echo transducer and a rotating mirror. For high resolution both in the axial as well as in lateral direction, the transducer must generate an extremely short acoustic pulse and have a wide radiation angle. A short acoustic pulse means an ultra wide band transducer. A wide radiation angle can be accomplished with a sharp focus. A sharp focus requires an aperture of the transducer which is large compared to the wavelength of the central frequency. Restricted by the maximum dimensions of the probe in the cervical channel, the maximum diameter of the ultrasonic wide band PVDF (Section 17.1) transducer is 2.5 mm (approximately 0.1

inch). The central frequency of the focussed transducer is 23 MHz. The bandwidth 135% (7.5–38 MHz at the –6 dB levels). At a central frequency of 23 MHz, the dimensions of the transducer is 20 times larger than the wavelength (a radius of 1.25 mm vs $\lambda \approx 60$ mm) and sharp focussing is possible. The lateral resolution of the focal zone is 110 mm and the axial resolution for pulse-echo measurements is 45 mm. The mirror rotates at a speed of 16 rotations per second while the axial position is controlled by a stepping motor. An angular encoder connected to the rotating motor, generates trigger pulses to the data acquisition unit. Within one revolution the probe records pulse-echo responses at up to 2048 different angular positions. The axial step size is 25 μm, which is approximately the wavelength of the maximum frequency. An accurate positioning of the acoustic beam inside the cervix is necessary for high resolution imaging.

Besides the evacuation of thyroid cysts under US guidance, ethanol sclerotherapy is a new therapeutic modality in cystic and solid nodular diseases. Percutaneous ethanol injection sclerotherapy (PEI) requires ethanol instillation into a nodule, which must be performed under US guidance in order to avoid damage to the extranodular parenchyma. PEI treatment is a promising new tool, but its indications are still debated and it requires great skill from the operator. It seems best at present to regard PEI treatment as an investigational method.

An experiment was conducted to check the efficacy of ultrasonography. Patients suffering from hepatocellular carcinoma (HCC) underwent enhanced ultrasonography with intravenous ultrasonographic agents Levovist, after transcatheter arterial embolisation (TAE) (size range 23–35 mm, mean 25.3 mm) or PEI treatment (15–35 mm, mean 24.5 mm), to compare the therapeutic effect with dynamic CT. The evaluation of enhanced ultrasonography correlated almost completely with that of dynamic CT. The residual lesions that could not be detected by dynamic CT due to lipiodol, were shown up clearly by enhanced ultrasonography and dynamic MRI. In total, eight residual tumours (3–15mm, mean 8.3mm) were seen. PEI treatment was perfomed on seven lesions and good control was obtained for six months. It can be concluded that enhanced ultrasonography with Levovist can be performed to evaluate the therapeutic effect of treatment for HCC, and can serve as a good marker for additional US-guided therapy.

17.2.2 Therapeutic USG

For therapeutic (e.g., lumbago, frozen shoulder) pain relieving applications, the systems operate at US power levels of up to many W/cm^2

A competitor for pain relieving, infrared rays possess an electromagnetic wave spectrum of 7600–15000 Å. These rays exert their effect on tissues by producing thermal energy. Erythematous rash relevant with the application of infrared rays is due to its stimulatory effects on vasomotor nerves. It expedites both arterial and venous circulation. The local heating effect associated with its use results in increased tissue and cell metabolism, and increased local circulation. Infrared therapy helps scavenge cellular particles and metabolites produced by trauma and inflammatory reactions. It also augments the flow of blood cells, antibodies, and essential substances like O_2 and cellular nutrients. Therefore, it has both a protective and restorative effect on tissues.

Ultrasound rays, beside its thermal and physical effects, exhibit a series of chemical effects such as hydrolisation, oxidation and depolymerisation. The rays exert their primary and crucial effect on bony tissues and later have a moderate to mild effect on muscle, nerve, and lipid tissues in sequence. The use of ultrasound rays holds out some hope in the treatment of calcaneus spine.

Ultrasound rays, within therapeutic limits, increase cellular membrane permeability and cellular metabolism in blood cells. They also cause vasodilatation both in the arteries and veins. The neural tissue manifestations brought out by the application of ultrasonography are: analgesia, reduced action potentials, neural block and paralysis. This explains well the way how ultrasound abolishes painful muscle cramps. Above the therapeutic limits, ultrasound rays may induce osteoporosis and cause hemorrhagic infarcts in the trabecular bone. This problem may be overcome by reducing the application time of ultrasonography to within 10 minutes.

A total of 10 sessions of ultrasound therapy may be given to patients at a reduced dose of 2 watt/cm^2. The advantage of using ultrasound therapy over infrared rays is that ultrasound rays give rise to deeper thermal effects on tissues. Thus, it is superior to infrared therapy at diminishing the symptoms of the patient. Infrared rays are merely effective for superficial heating. The effects of ultrasound therapy, especially on bony tissues, are more pronounced than that of infrared rays. Better results may be achieved with the combined group compared to the infrared group.

17.3 Some Special Applications

17.3.1 Echocardiography

Majority of USG applications are pulse-echo based for detecting and locating abnormalities of structures within the body. Echocardiography (often called simply as 'Echo') and USG scanners use this mode by transmitting a train of short duration US pulses into the body and detecting the energy reflected by a surface or boundary separating two media of different specific acoustic impedances. It consists of a probe (transducer of piezoelectric crystal), PRF generator (which produces a train of pulses controlling the sequence of events and which varies the pulse between 0.5 and 3.0 kHz), transmitter, receiver, circuit for T–R matching, wide band amplifier, swept gain control, video amplifier, time delay unit, time base, time marker and CRT display. Short time USG, with the presently used standards, is rather safe biologically compared to radio-contrast CT scan or even plain X-rays.

17.3.2 Foetal Monitor

It is important for doctors to have a working knowledge of diagnostic ultrasonography, especially in the field of obstetrics. The following are the reasons:

- Need for clinical information at the time of patient contact in a remote setting.
- Urgent clinical problems requiring immediate assessment.
- Higher specificity of obstetric ultrasonographic information obtained by clinicians who know the patient.
- Significant reductions in time and cost, which are important elements in managed care practices, and thus affect utilisation and quality.
- Desire for comprehensive obstetric skills by many family physicians.
- It is cost-effective in family practices where more than 100 patients are delivered per year, since the cost of equipment and development of office skills can be easily justified.

The impact of incorporating this skill into practice has been, and will continue to be, significant:

- There is immediate clinical correlation that assists in expedient patient management.
- There is improved continuity of care.

- It allows time for an educational interaction between patient and provider.
- It results in improved patient access.
- It expands the skills of family physicians.
- Driven by the generalist market, the industry will continue to expand its efforts in technology automation, cost reduction and market development.
- Incorporation of these skills into practice provides a direct benefit under managed care capitation to consumer, intermediary user and provider.

Scope of practice

Diagnostic ultrasonographic examination by family physicians is an appropriate skill, which enhances the diagnostic and therapeutic capabilities of generalist physicians. There is a wealth of information regarding the varied application of this intervention in family medicine. These interventions can be divided into the following general areas:

- Diagnostic ultrasound scans as a part of first-trimester pregnancy care.
- Diagnostic ultrasound scans as a part of second- or third-trimester pregnancy care.
- Gynecologic scans.
- Emergency scans performed in acutely ill patients during labour and delivery.
- General abdominal scans.
- Ultrasound-guided procedures (i.e., amniocentesis, external cephalic version, paracentesis, thoracentesis, organ biopsy, mass biopsy, dilation and curettage, intrauterine contraceptive device retrieval and insertion).
- Ultrasonographic scans of specific small parts and surface anatomy.
- Cardiovascular scans.

17.3.3 Laser Doppler Blood Flow Meter

In 1842, Christian Johann Doppler for the first time described the Doppler effect he discovered. The Doppler effect is based on the principle that the change of the distance between the observer and the source of sound is always associated with a physical change of the sound's frequency to the observer. Let us underline this effect by an example: if we stand on the road and an ambulance with its siren blaring passes by, we will observe a sudden change in the sound of the siren. Please note that we are now talking about the pitch, not about the intensity of the sound. The nearer the ambulance comes, the louder is the sound, which will then gradually decrease after the ambulance has passed. In contrast to the intensity, the pitch will remain constant the whole time. If the ambulance drives at a speed of 100 km/h, the pitch will vary in a range that covers two whole tones. The siren itself always blares at the same pitch but the frequency that is received by the human ear is influenced by the fact whether the sound comes towards us or moves away. The same principle can be applied if we pass in our own car, an ambulance that stands with a blaring siren (transmitter at fixed position with the receiver moving) or when both we and the ambulance are moving (transmitter and receiver in relative motion). When the transmitter and the receiver approach each other, the frequency received by the receiver lies above the transmitted frequency, while the frequency received is below the transmitted frequency, when the transmitter and the receiver move away from each other. The difference between the transmitted and received frequency is referred to as the *Doppler frequency* or *Doppler shift*.

In medical diagnostics, the Doppler effect is normally used for measuring the blood flow velocity in vessels. Foetal monitors measure the foetal heart rate (FHR) by the Doppler shift of foetal blood flow with the help of narrow beam or wide-angle beam of ultrasound waves. It was in 1959 that Saturmora first described how the Doppler effect could be used to measure blood flow velocity.

The transmitted ultrasound wave is reflected by the red cells (erythrocytes) in the blood stream. The frequency received by the transducer differs from the frequency that has originally been send into the body. In this case, with the transmitter and the receiver at fixed positions, the change in frequency is caused by the motion of the reflecting structure (red blood cells). This effect leads to a double shift in frequency — the first frequency shift taking place when the ultrasound wave meets the reflector and the second shift caused by the deflection of these echoes with their already changed frequency. Another factor that has an influence on the degree of the frequency shift is the angle between the vascular axis and the ultrasound beam, which is referred to as the *Doppler angle.*

The Doppler equation is:

$$\Delta f = \frac{v 2 f \cos\alpha}{c}$$

where: Δf = Doppler frequency
f = transmitted frequency (transducer)
c = velocity of sound
v = velocity of blood flow (relative target velocity)
α = angle between the vascular axis and the ultrasound beam

After the transformation of the equation to v, the velocity of the red blood cells can be calculated by the following equation:

$$v = \frac{\Delta f c}{2 f \cos\alpha}$$

The Doppler shift in ultrasound systems falls within the audio frequency range. The continuous wave (CW) Doppler instrument transmits a beam of 3–10 MHz US towards the blood vessel. Range gated pulsed Doppler flow meter comprises a transmitter (2–10 MHz) – a master oscillator driving the pulse repetition frequency (PRF) ripple counter and a receiver (band width of about 3 MHz and gain > 80 dB).

In contrast to continuous wave Doppler systems, pulsed wave Doppler systems are not able to generate a continuous Doppler signal because they do not continuously transmit and receive ultrasonic waves. In pulsed wave Doppler systems, the Doppler signals are generated on the basis of different measurement points, the so-called sample volumes. As the temporal resolution is given by the intervals between the transmitted pulses, which is reciprocal to the pulse repetition frequency, the actual propagation of the beam can not be determined. Therefore, the values measured within the sample volumes are interpolated and displayed on the screen. The upper limit of the frequency that can be measured has to be lower than half the pulse repetition frequency. Higher frequencies can not be represented because of missing measurement values. Therefore, the pulse repetition frequency has to be set twice as high as the Doppler frequency that should be measured.

Spectrum analysers can measure power spectra (FFT) and Doppler signals off-line. Laser Doppler blood flow meters emit light from a low power (5 mW) He—Ne laser coupled to a quartz fibre. Two beams (reference from the non-moving tissues and Doppler-shifted from the moving rbcs) are received by a plastic fibre and are transmitted back to the photo-diode for optical heterodyning. From this, the output velocity can be measured.

The frequency changes are converted to audio signals. Occasionally, simultaneous blood pressure (BP) measurement is carried out with a sphygmomanometer, to give the following index.

Pressure index (PI)

The ratio between the BPs measured in the posterior tibial artery at the ankle and at the brachial artery. It is usually greater than 1.0. If it is less than 0.9, it implies that there is some amount of arterial obstruction. The values between 0.9 and 1.0 are equivocal.

Colour coded Doppler (CCD)

This is a tool which allows a qualitative investigation of vascularisation, because structures which move differently in the organism are imaged in differently coloured forms. The colours themselves have no anatomic relevance. Flow toward the transducer is displayed as red, whereas blue indicates flow in the reverse direction. CCD is performed with a high-resolution linear transducer in the range from 5 MHz to 10 MHz. Correct setting of the wall filter and pulse repetitive frequency (PRF) is very important. The parameters most frequently used are as follows: a reference of 5 MHz, a wall filter of 50 or 100 Hz, and a PRF of 1500 Hz. Colour sensitivity is optimised by raising the colour gains to just below the threshold of colour noise. The output power is the sonographic energy applied to the patient. The lowest power is advisable, because CCD intensity levels may be a multiple of those with 2D-sonographic equipment. Pulsed wave Doppler systems are able to derive Doppler signals for exactly defined structures by means of a fixed measurement distance, the so-called sample volume. For the simultaneous analysis of the flow velocity in different structures, various different sample volumes are placed along the beam axis of several beams that run parallel to each other. Thus, the flow velocity can be analysed for an exactly defined area which is referred to as the *region of interest (ROI)*. The simultaneous spectral analysis for the different sample volumes by means of the fast Fourier transformation would take too much time and the simultaneous representation of a complete Doppler spectrum that may consist of 20 beams and 50 sample volumes per beam, i.e. 1,000 measurement points, would be impossible. This is why in colour Doppler mode, auto-correlation procedures are used for signal transformation.

To this purpose, all Doppler signals are collected for all relevant sample volumes and will later on be compared to the data of the following scan. While the analysis of the Doppler signals on the basis of the FFT spectral analysis requires at least 128 measurement points, auto-correlation procedures are able to calculate the Doppler shift on the basis of only 4 measurement points. From the measured Doppler shift, the average velocity within the sample volume can be derived. The measured values are represented on the B-Mode image on the screen in different colours. Blood flows moving towards the transducer are indicated in red, and blood flows moving away from the transducer are represented in blue. The velocity of the blood flows is indicated by points in different brightness – bright pixels stand for high velocities while low velocities are indicated by darker pixels. Angle-dependency and aliasing of the colour Doppler mode are as in pulsed wave Doppler mode. The aliasing effect in colour Doppler images is indicated by a change of colours.

Modern ultrasound systems include an additional mode by means of which the amplitudes of the returning echoes can be encoded and displayed in a colour map. This new mode has been described in 1994 for the first time and is known as: *colour-flow angio-mode, power Doppler mode, power Doppler angio-mode, colour angio-mode, angio-Farbe, colour Doppler energy or colour perfusion imaging.*

In colour Doppler mode, high-pass filters are used to distinguish the echoes returning from the tissue structures from those that are reflected by blood corpuscles (see Section 9.6). With the newly developed colour angio-mode, the amplitudes (intensities) of the returning echoes can be displayed. As the amplitudes of echoes returning from the tissue structure are often 1000 times higher than the intensity of signals reflected by erythrocytes, the signals can easily be distinguished from each other. In colour Doppler mode, the echoes returned from the blood are analysed by frequency analysis. In colour angio-mode, the amplitudes are encoded and displayed according to the same principle as in the B-scan mode which means that the echoes are displayed in specific colours according to their intensity.

As only the intensity of the echoes returned from the erythrocytes are analysed and the frequency shifts are not displayed, no aliasing effects or colour shifts resulting from the change of direction will appear on the colour flow images. On the other hand, it is no longer possible to distinguish between different velocities or blood flow directions on the image. With the colour angio-mode very low velocities can displayed. Further advantages are its low angle dependence and its excellent signal-noise ratio.

17.4 Further Reading

[1] Raja Rao, C. and Guha, S. K. *Principles of Medical Electronics and Biomedical Instrumentation.* Universities Press India Co. Ltd. 2001.

[2] http://www2.umdnj.edu/~shindler/echo.html

[3] www.hitachiultrasound.com

17.5 Questions

1. Descriptive Questions

1. What are the advantages and disadvantages of USG as a clinical diagnostic modality?
2. How does the Doppler effect help a clinician obtain an useful image ?

2. Write Short Notes on

1. CW Doppler
2. PW Doppler
3. Laser Doppler blood flow meter
4. Foetal monitor
5. Echo cardiogram

3. Multiple Choice Questions

Choose the single best response.

1. Ultrasound waves are above
   (A) 20 Hz
   (B) 20 kHz
   (C) both
   (D) none

2. Piezoelectric crystals do not

(A) mechanically deform in an electrical field
(B) generate an electric potential difference when deformed mechanically
(C) act as an ultrasound generator or transmitter
(D) suffice alone for major medical applications

3. A mode scan may be used for
   (A) determining crown rump length (CRL) of a foetus
   (B) abdominal mass
   (C) heart valves movement
   (D) any of the above
4. The thermal effect of US of raising temperature of tissues is due to
   (A) high absorptivity of tissue
   (B) low thermal conductivity of tissue
   (C) both
   (D) none
5. Therapeutic ultrasound is mainly used for
   (A) relief of pain
   (B) relief of swelling
   (C) relief of tumours
   (D) relief of malignancy

Answers

1. B; 2. D; 3. A; 4. C; 5. A

18

X-rays, CT, MRI and other scanners

18.1 Introduction

What was earlier known as radiology and radiodiagnosis is now better known as imagology. As far as medical imaging is concerned, there are many image processing techniques involved, some of which are discussed in Chapter 6. However, going into the details of image processing is beyond the scope of this book. One important aspect of medical imaging is 3-D (three-dimensional) reconstruction. The common uses of 3-D image reconstruction are (i) Scanning or transmission electron microscopy (SEM /TEM) e.g., of protein structure, (ii) CT (Surface rendered or MIP i.e., maximum intensity projection) and (iii) MRI.

Nuclear imaging modalities prepare cross-sectional images by detecting the distribution of radioactive particles (e.g., radio isotopes administered intravenously or intranasally. The radioisotopes emit γ-ray photons or positrons, which are the products of nuclear decomposition. Their concentration in any cross-section changes temporally with radioactive decay, flow and biochemical kinetics within the body. The utilisation of these properties help in the emission imaging armamentarium presented in Section 18.5.1 and 18.5.2.

To understand magnetic resonance, we must first learn about magnetic moments in any material. Nuclei of any cell have either protons or neutrons or both. If the number of protons and/or neutrons is odd, there will be a nuclear spin and a magnetic moment (M_0) having both magnitude and direction. Normally, in the living tissues of the body, the magnetic moments of the nuclei of the cells in the tissue are aligned randomly, giving rise to a net magnetisation of zero. These features are elaborated in Section18.4.

In accordance with the electromagnetic theory, any nucleus with a magnetic moment (e.g., hydrogen) tries to align itself with the magnetic field where it is placed. This leads to precession or wobbling of the magnetic moment about the applied magnetic field with a resonant angular (Larmor) frequency, denoted by ω_0 or $W = 2\pi v = rB_0$ or γB_0, where r or γ is a constant called *magnetogyric* or *gyromagnetic ratio* or the ratio between the magnetic moment and the angular momentum, and B_0 is the strength of the applied magnetic field or the density of the steady magnetic field and v is the frequency of radiation.

18.2 X-rays

X-rays are electromagnetic radiations of short wavelength and high energy which can penetrate materials that readily absorb and reflect visible light. For medical diagnostic radiography, the X-rays used are of the 1–10 nm wavelength range. In stationary anode tubes, the heat generated due to collisions of electrons at high speed, is usually dissipated onto insulating oil by radiation. These are employed in smaller X-ray machines. In rotating anode tubes, the target is replaced by a cooler one when the temperature due to electron bombardment becomes very high. For diagnostic purposes, high current (mA) and low exposure time is required and for radiotherapy, high voltage (kV) and relatively lower mA are required.

With the advent of the newer diagnostic modalities described in the subsequent sections, conventional radiography has taken a backseat. However, often a plain (without radio opaque contrast material) chest radiograph (CXR) or an abdominal X-ray is helpful in emergency conditions like those found in rural or backward settings.

Mammography is a soft tissue X-ray, taken by placing the breast (for detecting cysts, either benign or malignant) in direct contact with an ultra-sensitive film and exposing it to X-rays with low voltage and high amperage.

Xeromammography or *xerography* is an alternative X-ray technique using a photoconductor producing the final image on paper instead of film.

Other special types of X-ray include the barium swallow, meal and follow through for respectively esophageal, gastric and intestinal disorders and intravenous urography (IVU or IVP for IV pyelography) for renal, ureteric, urinary bladder and urethral disorders. Nowadays instead of IVU, DTPA (diethylene triamine penta-acetate) is preferred.

Fluoroscopy is a real-time X-ray used for guiding the introduction of endoscopes or catheters in blood vessels.

Digital subtraction angiography (DSA) is an invasive method which includes the direct intrarterial injection of a radio-opaque dye like Niopam (or occasionally to avoid the hazard of IA injection, IV radio-opaque solution is given) into the blood vessels to note their potency (e.g., in the heart: coronary angiography or kidney: renal angiography). The X-ray images are digitally subtracted to give an idea of the size of the arterial lumen in question.

18.3 Computed Tomography (CT) scan

Computed tomography (CT scan) or computerised axial tomography (CAT scan) is an operator-independent imaging modality. The patient lies on a movable table with the head (or any other region of interest) positioned in the centre of a larger circular frame or 'gantry'. The gantry contains an X-ray tube, which can be rotated 360° around the patient. Lining the gantry tunnel, at equally spaced positions are hundreds of detectors that receive the X-rays as they pass through the body. Likewise, a series of X-ray measurements are made from all possible orientations around the patient, usually in transverse sections (e.g., 8, 9 or 10 mm apart). When all measurements are complete, the information is processed by an adjoining computer and the X-ray pictures for any section of the body are reconstructed, as long as the plane of view is perpendicular to the mid-line (sagittal and frontal or coronal) axes of the body. Usually, 10–15 images are examined during a CT scan of the brain, taking 15–30 minutes. Each cross-sectional image, however, takes about 2 seconds. Apart from the injection of the IV (intravenous) contrast dye, CT scan is non-invasive. The level of radiation exposure is kept well within safe limits for diagnostic work. Typically, it can detect lesions or tumours as small as 3–4 mm^2 in area. Now spiral, 3D-CT and colour CT scanners are also available.

The basic components of a CT scanner are:

1. **Scanning system:** includes X-ray sources and detectors

2. **Processing unit:** It deals with the following information:

   i. *Positional:* like for example which traverse is being performed and how far the scanning frame is along its traverse
   ii. *Absorption:* values of attenuation coefficient from the detectors
   iii. *Reference:* data from the reference detector that monitors the X-ray output
   iv. *Calibration:* obtained at the end of each traverse

The first stage of computation involves analysis and conversion of collected data into a set of profiles, usually about 180 or more. The chief work is to reconstruct this data into meaningful pictures. The main reconstruction techniques are:

i. *Back projection* which is similar to graphic reconstruction
ii. *Iterative methods* which implement algebraic solutions
iii. *Analytical methods* which use exact formulae. Two common processes are filtered-back projection (through convolution of data) and Fourier (2-dimensional) filtering reconstruction techniques.

3. Viewing system: For viewing, both CRT and photographic plates are used. The absorption values are assigned a linear scale (e.g., in an original EMI scale, air is taken as –500, water as zero and bone as +500). Gray values and Hounsfield scales are also used.

4. Storage unit: For storage and subsequent analysis, magnetic disk tape or floppy disk, as well as photographs are used.

The image quality of a CT scanner depends on the high spatial resolution, low contrast resolution and rejection of noise and artifacts.

Conventional slices of CT scan images vary from 5–10 mm; while, contiguous CT/MRI images are taken as 1–2 mm slices.

Common applications of 3-D reconstruction of CT scan images include (i) tumour removal, (ii) cosmetic surgery; (iii) radiotherapy planning.

There are two types of 3-D reconstruction of CT scan images – surface modelling and solid modelling (MIP or maximum intensity projection). Surface modelling reconstructs the object as a shell. Structures are digitised manually or automatically and tissue edges are represented as streams of coordinates or contours. A viewing space is defined by mapping a window and its world coordinate contents into the viewport through a scaling function. The scaling function consists of the magnification factor and tissue thickness values. Geometric mapping algorithms are used to map coordinates for each contour to the subsequent contours on subsequent sections providing a wire-mesh model. The mesh model is rotated to a viewing angle and each surface tile is shaded according to the position of the light source and the viewer's eye. 3-D surface modelling may use polygon-based mesh models created from CT images in DICOM or other image formats. Three-dimensional models can be used for 3-D measurement and volume calculation, surgical and treatment planning, surgical simulation and 3-D rapid prototyping applications. Volume rendering uses either the entire image volume, a portion defined by regions of interest (ROI), or a portion defined by object boundaries. Object boundaries may be used to create 3-D mesh models and generate quantitative measurement reports. Able Software Corp. 3D-DOCTOR (Lexington, MA) is a 3-D imaging software using vector-based surface model creation rather than raster-based systems to significantly improve computational efficiency and the quality of generated 3-D models to be used by 3-D printers and rapid prototyping machines. In solid modelling, voxel stands for pixel volume. Registration is the determination of a one-to-one mapping or transformation between the coordinates in one space and those in another, such that points in the two spaces that correspond to the same anatomical point are mapped to each other. The readers may see <http://www.surfdriver.com/docs/reconstruction.html> for further details.

Solid modelling (MIP) for 3-D reconstruction of CT images considers the object as a discrete volume viewing space and the data input as voxels stacked upon each other. Here the intensity is computed, applied to the stack and rotated to give the final view.

18.4 Magnetic Resonance Imaging (MRI) Scan

Magnetic resonance imaging (MRI), formerly called nuclear magnetic resonance or (NMR), is based (not on X-radiation) on the combined use of a magnetic field, radio waves (radio frequency or RF with wavelength between 1 to 10^5 m) and a particular character of H atoms which are present in all tissues of the body. Usually, the nuclei of these atoms spin like tops, in random directions. When the MRI subject is positioned in a magnetic field, the nuclear spins of all the hydrogen atoms become oriented in a single direction. The scanner then delivers pulses of RF waves at these nuclei, temporarily jarring them out of alignment. When the nuclei rebound, several milliseconds later, they emit faint RF signals that can be interpreted by a computer. As the emitted energy varies in accordance with the density of the tissue, the computer is eventually able to construct an image in sagittal, frontal and axial planes.

Sometimes an IV contrast dye like Gd-DTPA (Gadolinium-diethylene triamine penta-acetate) is also used. Albeit there is no radiation exposure (usually contrast dye is not required), spatial details are superior to those in CT scans, images can be reconstructed in any plane and there is no chance of arachnoiditis (unlike in myelography), there are some disadvantages too. Apart from the prohibitive cost, because of the longer time necessary, it is not suitable for acute emergency. There is a greater chance of movement artifacts like 'blurring' and 'ghosting'. It also cannot be used in subjects having artificial cardiac pacemakers or metallic orthotics and prosthetics (because of magnetism). It is also insensitive to calcium and cortical bones. However, it is an excellent procedure for distinguishing white matter from gray matter and cerbro spinal fluid (csf) and also for visualising the spinal cord and canal. The strength of magnetism in MRI is about 1 tesla (1T or 10^{15} fT, i.e., femtoTesla), while the strength of the steady magnetic field of the earth is about 10^{11} fT or 10^{-4}T. In the case of MCG (magnetocardiography), the strength is about 10^5 fT. The cerebral a-rhythm is about 10^3 fT and cerebral EP about 10^2 fT, whereas the noise within a shielded room is only about 1 fT. The present day magnetometer can sense about 5 fT / √Hz.

At room temperature, more protons are in a low energy state than in a high energy state. During MRI, excited protons tend to return to its low energy state with spontaneous decay and re-emissions of energy as radio-wave photons at a later time *t*. This exponential decay produces a free induction decay (FID) signal that is fundamentally obtained from MRI systems.

If in a static field, RF waves of the right frequency are passed through the tissue in question, some of the protons in the parallel direction will absorb energy and be excited to a higher energy level in the anti-parallel direction. After some time, the absorbed RF will be emitted as EM energy of the same frequency as the RF source. The magnitude of energy necessary to flip protons from the parallel to the opposite direction is directly related to the magnetic field strength. If the tissue is subjected to an external magnetic field, the net magnetisation will shift from the Z-axis (*B* magnetic field direction) by an angle α. The same effect may be obtained with the application of an RF pulse at the resonant frequency of the tissue protons. The rotational angle a depends on the amplitude and length of the applied RF pulse. The RF pulse which can rotate the M_0 through 90° is known as the *90° RF pulse*. The net magnetic moment *M* also now precesses with the same characteristic frequency ω_0, since the individual magnetic moments causing this shift are in phase with the applied radio-waves.

When the RF pulse is turned off, the net tissue magnetisation begins to swing back towards the Z-axis, inducing an MRI signal in the receiver coil placed perpendicular to the moving magnetic vector. Now, individual magnetic moments start dephasing each other, some rotating faster and the others slower than the resonant frequency (ω_0). The component of *M* in the XY-plane disappears as the individual M_0 components in that plane cancel each other. The amplitude *A* of the initial signal received by the coil is proportional to *M* in the XY-plane (M_{xy}) and other geometric factors. *A* decays exponentially with time *t* as: $A = A_0 e^{-t/T_2}$, where T_2 is the average decay time for the process and the reciprocal of that is the decay constant.

Along with the dephasing decay process, there is relaxation of the M_z component to the resting state of M_0. This is also exponential with the average decay time of T_1. If T_1 is sufficiently short, the resultant signal amplitude with both the decay processes contributing to the observed decay signal would be: $A = A_0 e^{-t/T_1} e^{-t/T_1}$ Please note that the two relaxation processes are associated with excited nuclear spins.

T_1 or *spin-lattice* or *longitudinal relaxation time* denotes the time taken for the perturbed nuclei to realign themselves with the existing lattice structure of the host tissue. This time constant describes the recovery of the M_z component to the resting state of M_0, along the direction of the applied magnetic field.

T_2 or spin-spin or transverse relaxation time denotes the time taken for the perturbed, in-phase spins to dephase with each other. This time constant describes the decay of the M_{xy} component and is in the perpendicular direction to the applied magnetic field.

Transverse relaxation is faster than longitudinal relaxation; therefore, T_2 is always shorter than T_1. Both T_1 and T_2 are sensitive to the molecular structure and nuclear environment. T_2 tends to increase with increased nuclear motion, in fact, almost a million times (microseconds to seconds) from solids to liquids. In biological systems, hydrogen typically has T_2 of 0.04 to 2 seconds. The magnitude T_1 of H is inversely related to the amount of motion having spectral components near the resonant frequency. It may be in milliseconds in liquids and months in solids. Hydrogen in biological tissues has a range of T_1 as 0.05 to 3 seconds.

During a disease, the differences in relaxation times between various tissue types may be 100% greater than the normal values. This purveys a powerful contrast for anatomical and histopathological discrimination. In other words, T_1, T_2 and proton densities of the various normal and abnormal tissues together make this imaging modality worthwhile.

For interpretation, the time domain FID signal has to be converted to frequency domain by Fourier transform (Section 6.3.1).

The *Bloch equation* purveys a phenomenological description of the nuclear magnetisation M_t in the presence of the applied external magnetic field B_t, which is the input signal or stimulus and M_t is the output signal or response. The equation is:

$$\dot{M} = \frac{dM}{dt}\gamma M_x B - \left[\frac{M_x i + M_y j}{T_2}\right] - \frac{(M_3 - M_0)}{T_1}k.$$

where γ is the gyromagnetic ratio (Section 18.1) which is a physical property of the specific isotopic nucleus. $_1H^1$ or proton has γ of about 4.26×10^7 Hz T^{-1} or 2.68×10^8 rad $S^{-1}T^{-1}$.

The common image reconstructing techniques in MRI include the sequential point method, the sequential line method, the sequential plane methods (both 2- and 3- dimensional zeugmatography).

Contrast of the image is based on relaxation rates and is usually achieved through saturation recovery, inversion recovery or spin-echo imaging techniques.

A comparison of an MRI scan with a CT scanner is presented in Table 18.1.

Table 18.1 Comparative advantages and disadvantages of CT and MRI

Factors	CT	MRI
Availability	easy	Rare
Radiation	++	–
Time taken	much less	much more
Cost	+	+++
Soft tissues	not clear	Well-resolved
IV contrast	necessary for blood vessels	–
Axial image	large radiation required	–
Claustrophobia	+	–
Restrictions	minimal	pacemaker, ferro-magnetic bodies and clips, metallic implants

18.5 Other Important Modalities

18.5.1 Positron emission tomography (PET) scan

PET scans provide a more in-depth picture of what is occurring in the particular region which is observed than a CT scan. As an illustration, a CT scan can determine the number, shape, size of cars observed from the roof of a high-rise building. While a PET scan will be able to determine how many engines of those cars are running! PET scan and single photon emission CT (SPECT) can determine regions of hypoperfusion in the body. Local or regional blood flow (RBF) can be measured by monitoring the distribution of radioactive tracers with a battery of scintillation detectors. Usually, functional brain studies use this method.

The common positron emitters are ^{15}O, ^{13}N, ^{11}C, with half-lives of respectively about 2, 10 and 20 minutes. Either of these is tagged with 2-deoxyglucose (glucose being the major brain metabolite). The emitted positron almost immediately combines with an electron and the two get neutralised by the emission of two γ-rays. The two γ-ray photons travel in almost opposite directions, penetrate the surrounding tissues and are detected by a circular array of (72 or more) detectors. The scintillation detectors are usually made up of cesium fluoride, having a short light decay time constant of around 5 nanoseconds, minimising random coincidences, which may generate spurious data. After this, a computer algorithm rapidly reconstructs the spatial distribution of the radioactivity within the subject, in a selected plane and the resultant image is displayed on a CRT, with colour coding proportional to the BF.

18.5.2 Single photon emission computed tomography (SPECT) scan

With SPECT, the resolution of the image obtained can be enhanced greatly. For brain or cerebral BF (CRBF), a local uptake of 2-deoxyglucose is a good index. If 2-deoxyglucose is labelled with a short half-life positron emitter like (radio-isotopes of fluorine or oxygen) ^{18}F, ^{11}O or ^{15}O, the SPECT scan can monitor its concentration in any part of the brain, through the intact skull, in living people.

The basic principle is similar to a PET scanner, described in the previous section, except that instead of positrons, photons are detected here.

18.5.3 Nuclear magnetic resonance spectroscopy (NMRS)

(Nuclear) magnetic resonance spectroscopy or (N)MRS can measure concentrations of ATP, phosphocreatine, inorganic phosphate (Pi) and intracellular pH in muscles. It is also used for metabolic studies of muscles disorders through chemical shifts (change in resonating frequency due to alterations in magnetic field by 'shielding currents') associated with distribution around adjacent atoms in a complex environment.

18.5.4 Magnetic resonance angiography (MRA)

MR angiography (MRA) is very effective for visualising the vascular system well, although costly. The principle for this methodology is as follows.

If the frequency of an external magnetic field is kept near the Larmor frequency, nuclear magnets (like H, in the water of blood plasma) can be rotated and their presence detected by the application of this externally applied magnetic field. A region of demagnetisation can also be made by disorienting the nuclear magnets with a RF field near the Larmor frequency. To measure the (blood) flow in a blood vessel (= angio), the blood vessel has to be placed in a uniform steady magnetic field of strength or density B_0. The magnetic field will align the randomly oriented H atoms to B_0 either in parallel or anti-parallel orientations. The alignment will occur exponentially with a time constant T_l called the *longitudinal relaxation time*. More nuclear magnets align parallel rather than anti-parallel to B_0.

Radionuclide scanning: Thallium^{201} and technetium^{99} radio-isotope scans are used for the heart and bones respectively. Gamma cameras are used for both types of scans. Radio-isotopes are coupled to specific antimyocardial or antiosteocytic monoclonal antibodies. The areas of the heart or bone damaged by disease or injury may be delineated by this technique.

18.5.5 Functional MRI (fMRI) scan

Reduced hemoglobin (Hb) in the blood becomes paramagnetic, and it alters the signals emitted. Therefore, the amount of oxygenated and deoxygenated Hb can be measured – purveying an index of BF. The scan using this index to measure BF is called *fMRI* or *functional MRI*. It presently has a resolution that is better than that of a PET scan. A mathematical model may be used to measure the signal fluctuations of the blood-oxygen level dependent (BOLD) response.

The logistic and tactical differences of the many disciplines that collaborate to carry out fMRI research was apparent at a meeting (the fourth annual *fMRI Experience* conference held on May 13 and 14, 2002, at the National Institutes of Health (NIH), in Bethesda, Maryland, USA, organised jointly by research fellows at the NIH and Institute of Psychiatry (IoP) in London, where experts in neuropsychology, neurology, physics, mathematics, electrophysiology, biology and cognitive psychology worked together to deliver presentations about the strengths and limitations of the technique, and its relevance in exploring the functional architecture of the healthy and disordered brain.

18.5.6 Thermography

Medical thermography is done with a sensitive infrared camera displaying a video image (in a CRT or also a static one on a photographic plate) of the temperature distribution over the surface of the skin. This image enables temperature differences to be visualised instantaneously, purveying evidences of abnormality, if any. It indicates the metabolic and circulatory (physiological or functional) changes, rather than the anatomical or structural ones. Our body absorbs irr (infrared radiation) almost without reflection and, simultaneously, emits part of its own thermal energy as irr. The intensity of this radiant energy depends on the temperature of the radiant (body) surface. So, it is possible to measure the varying intensity of radiation (surface temperature) at a distance from the body.

The advantages of thermography include the fact that it is an absolutely non-invasive, non-contact, radiation hazard free method and that images can be generated real time.

The infrared electromagnetic frequency is between 0.77 and 100 mm (higher than radio frequency but below visible light). The radiations are emitted spontaneously by any object at a temperature above absolute zero. The total energy W emitted is related to its temperature by the Stefan–Boltzman formula

$$W = \sigma\varepsilon T^4$$

where

$$\lambda_{max} = \frac{2897(\mu m)}{T(K)}$$

and $\lambda_{max} \approx 10\ \mu m$.

The irr increases with inflammatory and malignant conditions. However, the sensitivity (less false negative cases i.e., the number of malignant cases diagnosed as non-malignant) and specificity (less false positive i.e., the number of non-malignant cases diagnosed as malignant cases) of thermography is not high.

18.5.7 Fusion of these techniques

Hybrid or fusion technologies combining MEG/MRI/fMRI/PET scans are now available. Also see Chapters 2 and 6 for comparison.

A combination of MRI and MEG is now routinely used abroad. For pre-operative evaluation, 3D-reconstruction of the outer surface of the brain from an MRI may help the surgeon identify the brain structures after opening the skull, when the brain tissue retracts and the relationships change. Surface models which reconstruct the organ help better orientation during surgery.

Draulio de Araujo (University of Sao Paulo, Brazil) has presented results from his doctoral research, which has combined fMRI with MEG to examine the neural and electrophysiological basis of human navigation, using impressive virtual reality stimuli. Vincent Walsh (University of Oxford, UK) has also similarly described the utility and logistics of combining fMRI with transcranial magnetic stimulation (TMS) – which can be used to create a 'virtual lesion', or disrupt normal activity in cortical brain regions – as a means of determining whether a specific region is necessary to perform a particular cognitive function. These are two of the recent works presented at the fourth annual *fMRI Experience* conference held on May 13 and 14, 2002, at the National Institutes of Health (NIH), Bethesda. With the major focus of the conference upon fMRI, other imaging techniques such as SPECT, PET, or temporal measures of neural activity (qEEG) were not dealt with directly. As such, attempts were seldom made to differentially evaluate the merits of each of these techniques. Furthermore, potential problems such as (a) integrating regional activation with other information from genetic, temperamental, cultural differences, (b) understanding the distinction between overt structural abnormalities and brain dysfunction, and/or (c) ascertaining the meaning of blood-oxygen utilisation in terms of functional utility (i.e., excitation, inhibition) were not addressed [13].

Compared to PET and fMRI, MEG has better temporal resolution enabling monitoring of cortical dynamics in milliseconds. In hybrid technology, active brain regions may first be detected with PET/fMRI and then used as source locations in inverse solution of MEG data for revealing their temporal behaviour. However, neuronal synchronisations (as in MEG/EEG) do not always induce significant changes in mean neuronal firing rates or blood flow (BF) and metabolism. Therefore, MEG/EEG and fMRI/PET are actually complementary methods for studying the functions of the human brain. MEG applications include research in anaesthesia, epilepsy, ischemic lesions, migraine, ECT (electro-convulsive therapy or shock therapy for psychiatric disorders), encephalitis and Parkinsonism. Evoked potentials (EP) or evoked responses (ER) are possible (as in EEG) with auditory (AEP), visual (VEP) and somato-sensory (SSEP) stimuli in MEG.

18.6 Further Reading

[1] Edelman, R. R. and Warach, S. Magnetic resonance imaging. *New Engl. J. Med.* 328: 708. 1993.

[2] Haaga, J. R. and Alfidi, R. J., Eds. *Computed Tomography of the WholeBody*, 2nd ed. Mosby. 1988.

[3] Lee, J. K. T., Sagel, S. S., Stanley, R. J. and Heiken, J. P. *Computed Tomography with MRI Correlation,* 3rd ed. New York: Raven Press. 1995.

[4] Wegener, O. H. *Whole Body Computed Tomography*. Oxford: Blackwell. 1992.

[5] Felix, R. and Langer, M. *Advances in CT-II.* Berlin: Springer-Verlag. 1992.
[6] Pokieser, H. and Lechner, G. *Advances in CT-III.* Berlin: Springer-Verlag. 1994.
[7] Atlas, S. W. Ed. *Magnetic Resonance Imaging of the Brain and Spine*, 2nd ed. New York: Raven Press. 1996.
[8] Stark, D. D. and Bradley, W. G., Eds. *Magnetic Resonance Imaging*, 3rd ed. St. Louis: C V Mosby. 1996.
[9] Lopes da Silva, F. H. A critical review of clinical applications of topographic mapping of brain potentials. *J. clin. Neurophysiol.* 7: 535–551. 1990.
[10] American Academy of Neurology. Assessment: EEG brain mapping report of the therapeutic and technology subcommittee of the American Academy of Neurology. *Neurology* 39: 1100–1101. 1989.
[11] American Psychiatric Association. Quantitative electroencephalography: A report on the present state of computerized EEG techniques by the American Psychiatric Association Task Force on Quantitative Electrophysiological Assessment. *Am. J. Psychiatr.* 148: 961–964. 1991.
[12] John, E. R. The role of quantitative EEG topographic mapping or 'neurometrics' in the diagnosis of psychiatric and neurological disorders: The pros. *Electroencephalogr. clin. Neurophysiol.* 73: 2– 4. 1989.
[13] George, M.S., Ketter, T.A., Kimbrell, T.A., Speer, A.M., Lorberbaum, J., Liberatos, C.C., Nahas, Z, and Post, R.M. Neuroimaging approaches to the study of emotion. In, *The Neuropsychology of Emotion,* pp. 106–134. Ed. J.C. Borod. Oxford: Oxford University Press. 2000.
[14] Hari, R. and Salmelin, R. Human cortical rhythms: A neuromagnetic view through the skull. *Trends Neurosci*. 20: 44–49. 1997.
[15] Hashimoto, I., Okada, Y. C. and Ogawa, S., Eds. *Visualization of Information Processing in the Brain.* Amsterdam: Elsevier. 1996.

18.7 Questions

1. Descriptive Questions

1. Compare and contrast CT and MRI scans in terms of application.
2. What are the new imaging modalities (after CT and MRI) being developed for better assessment of medical conditions?

2. Write Short Notes on

1. CXR (chest X-ray)
2. IV contrast CT scan
3. MRA
4. PET
5. SPECT
6. fMRI
7. MRS
8. Radioisotopes used for PET scan
9. Gd-DTPA
10. CRBF

3. Multiple Choice Questions

Choose the single best response.

1. In SI units Tesla is expressed as:
   (A) $NA^{-1}m^{-1}$
   (B) $NA^{-1}m^{-3}$
   (C) $kgs^{-2}A^{-1}$
   (D) $kgm^{-2}s^{-2}A^{-1}$
2. Gadolinium (Gd) has a higher saturation magnetization than Co (cobalt) at
   (A) −273 °C
   (B) 250 °C
   (C) 290 °C
   (D) 796 °C
3. Transition from ferromagnetic to paramagnetic state is named after
   (A) Neel
   (B) Curie
   (C) Curie-Weiss
   (D) Debye
4. Field strength used in MRI is
   (A) 10.0 tesla
   (B) 7.0 tesla
   (C) 5.0 tesla
   (D) 1.0 tesla
5. Which of the following is not based upon electromagnetic radiations from our body?
   (A) EEG
   (B) MEG
   (C) X-rays
   (D) MRI

Answers

1. C; 2. A; 3. B; 4. D; 5. C

Section 7

Computers in Medicine and Healthcare

Computers are now ubiquitous in the medical and healthcare setup.
Medical or *healthcare informatics* is the sçience or art of processing medical and health related information. Biomedical informatics encompasses both healthcare informatics, as well as bioinformatics (often loosely termed as Computational Biology). Chapter 19 deals with diagnostic decision support systems, connectionist expert systems in medicine and health care and evidence based medicine.

19

Medical Informatics

19.1 Introduction

Medical (healthcare) informatics, a novel academic discipline, bridging medicine and information sciences, has already made its presence felt. It may be defined as the art and science of processing medical information.

The modern age can be correctly called the 'age of information' and every field of study is becoming so inundated with information that people are apprehensive of an 'information explosion'. Improper access to excessive information leads to nothing but confusion. This becomes truer in conditions of life and death – which the medical profession has to handle regularly.

Bioinformatics related to proteomics (informatics of protein structure banks), genomics (informatics of genomes and gene banks) and various new drug synthesis methods are not included in this book.

Information is derived from the assembly, analysis or summarising of data into a meaningful form. Data mean any group of operands or factors consisting of numbers (numerical), alphabetic characters or symbols (linguistic), which denote any condition, value or state. Knowledge comprises generalised truths formed from the analysis of available information. Medical or health informatics (dealing with medical information), is a necessary discipline especially in tropical countries like India since databases on tropical diseases are scanty. Facts and findings on any endemic disease need to be studied and recorded systematically (with the help of informatics tools). For example, it has been found that goitre in Sub-Himalayan regions is common not only because of less iodine in soil and water, but also due to the local food habit where the staple food includes legumes and cabbage that contain goitrogens (which inhibit the uptake of iodine in our body). Treatment for bronchial asthma in Central India includes antifilarials since endemic filariasis is rather common here. Inclusion of these findings in the dynamic knowledge base can enrich modern medicine. With the development of excellent satellite telecommunications in India, the prospects for medical informatics are brightening. However, it essentially requires a multidisciplinary interaction.

To cite an example, an elderly lady who presents with a fractured forearm due to a fall, may have other problems leading to the fall. She might be having increased frequency of urination because of diabetes or urinary tract infection (or benign enlargement of the prostate, in case of a male). She might have overlooked an obstacle while going to the toilet, in the night, due to failing eyesight (which again may have been caused by cataract or drugs), and fallen. The fall may also be due to increased or decreased blood pressure or blood sugar. Moreover, staying alone in the house could also be responsible for the fall, as there would have been no one to escort her to the toilet. There could also be psychological reasons such as depression or dementia, which might make the patient less careful than in usual circumstances. Therefore, a forearm fracture, in an aged lady, has a greater amount of 'history' than the same in a young lady.

The scope of medical informatics is immense as it forms the interface between the subject domain of medicine and the science and technology of computing. It requires essential inputs from cognitive sciences (including psychology), information and communication sciences (library and documentation), computer sciences (hardware and software), logic, statistics (including biostatistics and epidemiology), decision sciences, clinical and basic (pre- and para-clinical) medical sciences, health economics and policy-making as well as from medical ethics.

The broad range of activities includes:

(a) *Basic research* Model building, theory development and innovative experimentation. The aim is to build new knowledge from existing ones and to structure and assimilate the knowledge in terms of cause-process-effects. Standardisation of the approaches to reaching a decision is also required.

(b) *Applied research* Formal experimentation with proper evaluation.

(c) *Engineering* Specific tool building and development.

(d) *Implementation* Operation, maintenance, management, updating of the applications and training the users.

(e) *Planning and policy development* Role of information technology in health care delivery and medical education.

The career opportunities in this field are plentiful. This sort of a 'fused' or 'sandwich' discipline requires expertise from various (apparently unrelated) fields and can, therefore, offer career openings to a large number of people. Students with backgrounds in many diverse disciplines can contribute fruitfully.

The areas which desperately need people are:

(i) *Academic medical informatics* Research and development, i.e., for educational support. This area requires intelligent candidates, physicians or other health professionals who have

pursued formal training in medical informatics as their specialisation e.g., through biomedical engineering.

(ii) *Clinical, administrative and educational management* Traditional clinical jobs with an added condition of having a deep understanding of information technology. Students who have been trained in various informatics laboratories will be suitable.

(iii) *Operational services management.* This area requires people who can develop new systems which integrate with other programs and evolve into service activities. Long-term maintenance and support is absolutely essential for the sustainable development of an information system.

(iv) *Hospital Information Management* This area needs senior executive officers.

(v) *Corporate research and development.* Students trained in pharmaceutics/biotechnology can get jobs as 'biomedical informaticians' for research and development (R&D).

Need for an Academic Discipline

For the purpose of research, training and logistic or administrative issues, it may not be possible to integrate and collaborate successfully at an individual level. Corporate or government help is required. Many countries have felt this need and therefore, a number of academic units for medical informatics have already been established in many European countries and the USA.

The American Medical Informatics Association supports a scientific academy of elected fellows called the American College of Medical Informatics. They also hold a large annual conference called the Symposium of Computer Applications in Medical Care (SCAMC). There are also several scientific journals in the field. *Computers and Biomedical Research, Methods of Informatics in Medicine, Medical Decision Making, M.D. Computing, IEEE Transactions on Biomedical Engineering, IEEE Transactions on PAMI, Medical Engineering and Physics* are some of the notable ones.

India is globally connected to the Internet. The major public Indian networks include the Educational and Research Network (ERNET) of the department of Electronics (DOE); the Scientific and Industrial Research Network (SIRNET) managed by the Indian National Scientific Documentation Centre (INSDOC) for CSIR and the NICNET set up and managed by the National Informatics Centre (NIC). The Ministry of Human Resource Development has made available the INDEST Consortium for accessing numerous reputed journals online.

Nowadays, most important journals are available online, sometimes even before the hardcopy is available in the market. With ISDN connectivity being provided by the BSNL and VSAT connectivity being provided by ISRO, telecommunication activities are usefully increasing in all fields including the healthcare delivery sector. Isolated activities in this regard are being carried out in the various centres or schools for biomedical engineering, electrical/electronics/communication engineering and artificial intelligence research units. However, collaboration between all these units is essential.

In India, the Computer Society of India (CSI) has formed a Special Interest Group in Medical Informatics (SIGMI). The Indian Association of Medical Informatics (IAMI) is actively involved in spreading medical informatics awareness through its website (www.iamindia.org) and the online *Indian Journal of Medical Informatics.* There are also other organisations like the Medical Computer Society of India (MCSI) and the National Institute of Medical Informatics (NIMI) dealing with medical informatics.

19.2 Health Information Systems (HIS)

The health or hospital information system (HIS) is an integrated system encompassing the departmental systems (like radiology, pathology and pharmacy), as well as the clinical systems (like the intensive care unit or ICU). However, a proper definition is yet to evolve.

The database includes all sorts of records for each patient, including admission, evaluation, and management, transfer, and discharge information, along with supporting evidence.

Decision support systems (Section 19.4.1) may also be incorporated into it. It is the integration of global needs for all the healthcare users. It may be realised by interfacing a central system (as described in the next section) to multiple departmental (e.g., pharmacy and radiology) and clinical information systems (e.g., intensive or critical care units and nursing units).

Middleware (It is a technology for defining objects and creating interfaces between software systems, or a software that connects two otherwise separate applications.) in HIS may purvey solutions which integrate data for end users at the front, while hiding the multiplicity of the origins of the data at the background.

There are various standards that have been (and are being) developed for medical communication. Some of the notable ones include: ASTM (American Society for Testing and Materials), ASC X12N, DICOM (digital image and communication), HL7 (health level seven), IEEE P1157 (from the Institute of Electrical and Electronics Engineers, Inc.) MEDIX (medical data interchange standard), IEEE P1073 MIB (medical information bus), and NCPDP (National Council for Prescription Drug Programs). Some of these have already been dealt with in Section 3.3.

19.3 Computer Based Patient Records (CBPR or CPR)

Computer based patient records (CBPR or CPR) involves digitising all data related to a patient being treated in a hospital (along with modifications and updates) and storing it in an easily retrievable or searchable form. The database may include patient admission, transfer, discharge, evaluation and management, as well as, hospital and ambulatory care systems, and at times also CDSS. A plethora of nomenclature like EMR/EPR (electronic medical or patient record) /PMR (patient medical record) abounds in literature for similar entities. However, present day stress is on EHR (electronic health record) i.e., the whole data record of a person while in health and when unwell.

A simplified flowchart of a hospital CBPR or CPR is presented in Fig. 19.1. NHS (UK) has also given a classification for the different levels of EPR, and this is presented in Table 19.1.

Table 19.1 NHS (UK) classification of EPR levels:

Level	Description
Level 1	Patient administration system (PAS) – Medical records and medical secretary's case note tracking – Medical records, medical secretaries, wards bed management (real time ADT's: automatic admit/transfer/discharge notification) – Number of smaller clinical systems in place as standalone systems including cancer systems, child development systems
Level 2	Common PMI - integration with departmental systems: • Pathology • Radiology

- Theatres
- A&E
- Maternity
- Foetal medicine
- ECG
- Nuclear medicine
- Nurse information system

The ATM network and PCs throughout may be enabled access to all systems from any PC with appropriate authorisation.

Level 3 Reporting of results for pathology inpatients throughout.Nursing assessments and care plans are live in all ward areas.Pre-assessment – live Discharge letters produced by medical secretaries using integrated PAS functions.Discharge notification and GP connectivity including referrals, protocols and scanning of referral letters.

Level 4 Intranet implementation includes Medway Pharmacy guidelines.Internet and NHS access to the desktop.Full code of connection as per NHS guidelines.

Level 5 NICU monitoring equipment linked with the clinical information system e.g., CDSS

Level 6 Network capable of supporting full telemedicine and video conferencing.Telemedicine supporting remote MIU; PACS – business case Document imaging – A&E records integrated with A&E departmental system

Other considerations X400 messaging in place – able to exchange email with external and internal users.NWCS; CMMS – inpatients/outpatients/waiting listsData repository – oracle/business objects with clinical data – A&E and theatres.Intranet –guideline/ policies/procedures and communicationDiabetes project – IA – sharing of data with GP and centre.

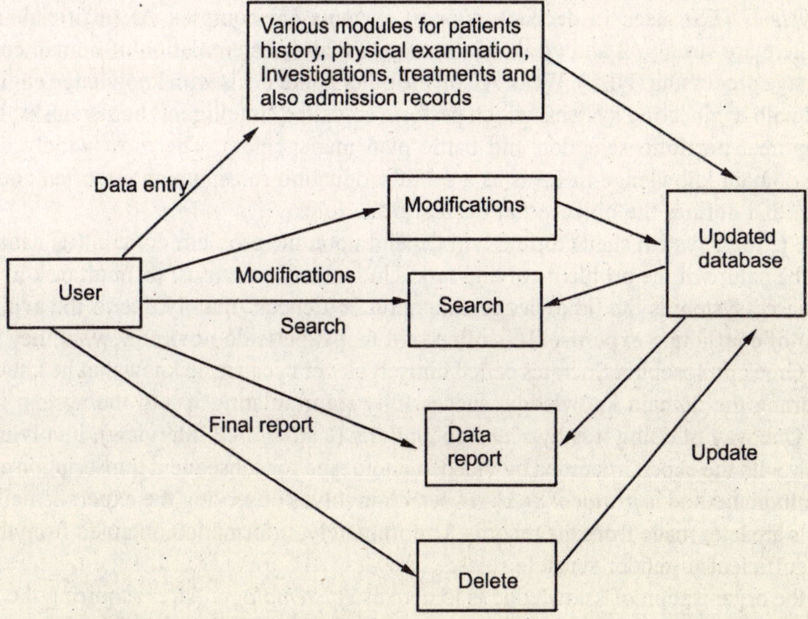

Fig.19.1 Flow chart for a simple CBPR

19.4 Clinical Diagnosis

19.4.1 Clinical or diagnostic decision support systems

Reaching a foolproof diagnosis is never an easy job for a clinician. Often, a simple diagnostic procedure or test is overlooked and the disease eludes diagnosis. Clinical reasoning and decision-making are phased. Initially there is a clinical evaluation (history taking and physical examination), followed by precise laboratory investigations. Then, integration of clinical findings and test results is done. After that, comparative benefits and risks are weighed among the alternative courses of actions, like drug interactions. Finally, the patient's preferences are taken into account, along with ethical and other considerations like cost of therapy, compliance expectations, and a therapeutic plan is developed. Right from the first step (history taking) to the final one, computers can be of immense help to the clinician.

While taking history from a patient, an experienced doctor does an 'iterative hypothesis testing', while a medical student or a fresh graduate learns to do 'comprehensive history taking'. The first method is known as the 'rule of the thumb' or heuristics, which comes from experience and is therefore bound to cause more errors despite saving time. As the human brain (and neural networks) works in milliseconds, computers can actually do a comprehensive history taking in a much shorter time.

The author has developed a few diagnostic decision support systems for medical education, research and communication. A simple example (using the iterative dichotomizer or ID decision tree) is provided in Appendix D.1. Despite the efficiency of computer-assisted diagnostic systems, a human clinician ('man in the loop' for 'intelligence amplification') is a necessary component of the system. Moreover, the clinician must understand completely the strengths and limitations of the system. Computerised diagnostics and clinical acumen are not mutually exclusive; rather they should reinforce each other for the alleviation of psychosomatic suffering of mankind.

Expert Systems (ES), used in decision support systems are complex AI (artificial intelligence) programs. There are strong AI and weak AI. Strong AI addresses emulation of human cognition like natural language processing (NLP). Weak AI, on the other hand, deals with knowledge engineering and is concerned with engineering systems which perform so-called 'intelligent' human tasks, like medical diagnosis, optimal portfolio selection and battle plan management. The most widely used way of representing domain knowledge in ES is as a set of production rules, which are often coupled with a frame system that defines the objects that occurs in the rules.

The frame (expert system shell) format will depend upon the problem domain (its nature and size) and also on the nature of the problem-solving task. Undoubtedly the most difficult task in designing a knowledge-based system is the 'knowledge elicitation bottleneck' mainly due to the availability of a huge amount of conflicting expertise. It is often said that experts do not know what they know. (The well-known Greek philosopher Socrates called himself clever because he knew that he knew nothing!).

After eliciting the domain knowledge, successfully reformulating it into the system is of utmost importance. One way of doing it is by *scenario analysis* (a structured interview), involving a face-to-face interview with the expert, recorded by video or audio tape for subsequent transcription and analysis. Another common method is *protocol analysis,* which involves observing the expert actually doing the job. Protocols are later made from the records. Unfortunately, information obtained from the protocols may not be sufficient to render sensible rules.

Study of the organisation of knowledge is known as *epistemology*. After acquiring the knowledge, an *epistemological analysis* is conducted, establishing the structural properties of the expertise.

Ultimately, a *logical analysis* is performed to map the knowledge to a formal structure. At this stage of the design process, the expert's knowledge is reformulated for the system. After the knowledge engineer codes the knowledge explicitly in the knowledge base, the subject expert (clinician) may then critically evaluate the system (repeatedly) until a satisfactory performance level is reached.

ES for medical diagnosis (clinical or diagnostic decision support systems)

In medical diagnosis, there is a huge scope for ambiguity in inputs, like, history (patient's description of the diseased condition), physical examinations (especially in cases of uncooperative or less intelligent patients), laboratory tests (faulty methods or equipment). Moreover, during treatment, there are chances of drug reactions and specific allergies, patient's non-compliance to therapy due to cost, non-availability of time or adverse reactions. Because of the almost daily advent of new treatment modalities, the process of making a decision about which particular treatment regime to adopt for an individual patient becomes very complex. More often, a large amount of information has to be processed, much of which is not quantifiable. Intuitive thought processes involve rapid unconscious data processing and combines diverse available information by law of average. Therefore, this has a low intra- and inter-person consistency. It is better if the clinician of today move towards analytic decision-making, which albeit typically slow, is conscious, consistent and clearly spells out the basis of the decision made.

19.4.2 Connectionist Expert Systems

The other branch of AI is ANN or artificial neural networks (please see Chapter 6). Because of their interconnections, the ANN approach is often called the connectionist approach. *Connectionist ES* are ANN-based ES where the ANN generates inferencing rules e.g., fuzzy-MLP where the linguistic and natural form of inputs are used. Apart from that, rough set theory may be used to encode knowledge in the weights better. Genetic algorithms (GAs) may be used to optimise the search solutions better. All these methods fall under the purview of '*soft computing*'.

For instance, a trained MLP [2] is used for rule generation in the If-Then form. These rules describe the extent to which a pattern belongs or does not belong to one of the classes in terms of antecedent and consequent clauses. For this, one has to backtrack along maximal weighted paths using the trained net and utilise its input and output activations.

After training, the connection weights of an MLP encode among themselves (in a distributed fashion) all the information learned about the input-output mapping. Therefore, any link weight with a large magnitude reflects a strong correlation between the connecting neurons. This property is utilised in evaluating the importance of an input node (*feature*) on an output layer mode (*decision*). For this, one needs to compute the path weights from each output node to each input node through the various hidden nodes. Each output-input path which is maximum (through any hidden node) denotes the importance of that input feature for arriving at the corresponding output decision. One or more hidden layers may be considered (if necessary) for evaluating the path weights. An increase in the number of hidden layers simply leads to an increase in the possible variety of paths being generated through the various hidden nodes in the different layers.

A heuristic thus allows selection of those *currently active* input neurons contributing *the most* to the final conclusion (among those lying along the maximum weighted paths to the output node j) as the clauses of the antecedent part of a rule. Hence, it enables the *currently active* test pattern inputs (current evidence) to influence the generated knowledge base (connection weights learned during training) in producing a rule to justify the *current* inference. The complete *If* part of the rule is found by *AND*ing the clauses corresponding to each of the features.

Because of the inherent fuzziness (qualitative nature) of most biomedical data, connectionist expert systems are likely to fare better than their traditional counterparts, as far as reaching a diagnosis is concerned.

Spatiotemporal contextual information can be incorporated for automated EEG analysis through syntactic analysis techniques. Here, EEG is represented as a series of elementary patterns called *tokens*. The EEG tracing is subdivided into 1-second intervals. A sequence of unique characters may be called a *label*. The resulting sequence of labels is usually known as a *sentence* which is 'parsed' by a definite grammar. A *grammar* consists of a set of rules that governs the merging of tokens into higher-order recognisable entities in the input data.

A severe drawback in this approach lies in the amount of heuristics involved in writing the grammar. Also, it is almost impossible for an untrained electroencephalographer (EEGer), to inspect the grammar and make any necessary modifications. Knowledge-based approaches purvey a way of increasing the involvement of the expert (EEGer) in the design process and also allow the utilisation of contextual information to a larger degree. These systems (ESs) apply a body of knowledge (knowledge base) to the input data and subsequently derive facts (data base). Typically, the knowledge base comprises IF<premise>THEN<action> type rules which presumably reflect human expertise. Another advantage of this type of flexible problem solving approach is that the collection of rules can be modified from time to time.

Knowledge-based enhancement of EEG signals and parallel processing are employed to save time and effort, as well as to increase the accuracy of the interpretations.

For details the interested reader may kindly refer to Sarbadhikari and Pal [2002].

19.5 Computers in Medical Training

Computer aided or assisted instruction (CAI) is a tutorial method using a computer as a base for managing the student's progress (at the desired pace and time). Computer aided or assisted learning (CAL) is a computer based tutorial method that uses the computer to pose questions, provide remedial information and chart a student through a course. Nowadays, with the emphasis, especially in medical related fields, on problem based learning (PBL), CAI and CAL have gained greater significance. In PBL, the approach is top down, rather than bottom up, i.e., instead of starting by learning physiology or biochemistry, one discusses a patient's presenting features and then goes backwards to find the physiological and biochemical basis of the disease presented.

19.5.1 CAL

Requirements for both CAI and CAL are the following.

(i) For using the available packages:

1. Interactive CDs on various topics.
2. LCD projector for addressing a large number of students simultaneously.
3. Proper orientation of teachers in all departments for full utilisation of the facility.
4. Devoting time slots (like tutorials) for using the facilities
5. Availability of such facilities in the students' hostel campus too.

(ii) For developing packages:

1. Selecting clinical cases with their photographs, radiographs, along with the physiological, biochemical, pathological and microbiological reports. These can be collected from the CBPR (Section 19.3).

2. Formulation of diagnosis, and then linking the terminologies with the respective subject databases. This would be also an example of CBR (Section 19.6.2).
3. Discussion of possible treatments e.g., exercise, diet, medicine, surgery, radiation – including drug-drug and drug-food interactions, cost and mode of therapy and possible compliance.
4. Further questions for self-assessment and possible answers from respective databases.
5. Scoring of responses from the learner.

Some common languages and interfaces for CAI and CAL are listed below.

1. MeSH (medical subject headings) classification system
2. UMLS (unified medical language system) from the NLM (National Library of Medicine, National Institutes of Health, Bethesda, Maryland, USA).
3. CBL demo (asymetrix toolbook's computer based learning demo)
4. CALScribe (ETS, University of Bristol, UK)
5. CTICM (CTI Centre for Medicine Template, University of Bristol, UK)

19.5.2 CAI

For all practical purposes, the CAI programs are insufficient alternatives to textbooks. Therefore, only the more comprehensive CAL programs are used commercially and academically. In CAI programs the process of learning through solving problems (making errors and corrections) is not present.

Video materials can supplement the CAI and CAL programs, for better understanding.

19.6 Newer Trends

19.6.1 Evidence based medicine (EBM)

There is no definitive method for selecting the appropriate alternative regime for a particular patient. Data sheets on various medicines purveyed by the pharmaceutical manufacturers are not of much help since they focus on only one particular (generic or brand) drug, while doctors require comparative data on all available treatment modalities. To overcome this problem, evidence based medicine (EBM) compiles data based on systematic review of RCTs (randomised control trials), through meta-analysis, involving all available therapies. To cite a concrete example, the comparative effectiveness of various interventions in relieving or reducing pain of osteoarthrosis (OA) can be summarized as: beneficial and likely to be beneficial. In the first group we get systemic simple analgesics like paracetamol (acetaminophen) for short-term pain relief and improvement in functions, systemic NSAIDs (non-steroidal anti-inflammatory drugs), which also give short-term pain relief and functional improvement and topical analgesic agents (for short-term pain relief). In the second group we find education, dietary advice, empowerment, and support (improved knowledge of the disease and pain relief). Physical support also purveys pain relief and functional benefits. The key point noted was that there was no good evidence that NSAIDs were superior to paracetamol nor was there anything to suggest that any one of the various NSAIDs available in the market has greater efficacy in pain reduction in OA. One systematic review of RCT has found that topical agents relieve pain in OA and offer a less toxic alternative to systemic drug therapy. However, there is no evidence to the effect that prescribed local analgesics are superior to cheaper, non-prescription OTC (over the counter) alternatives or local hot/cold packs. On recovery from active inflammation, a program of stepwise simple (isotonic) exercise (like quadriceps exercise or stretching and relaxing the knee joints in knee OA) is often essential.

19.6.2 Case based reasoning (CBR)

A medical practitioner encountering a new problem is usually reminded of similar cases which he might have seen in the past. New problems are solved by analogy with old ones and explanations and reasoning are often derived from prior experiences. This sort of problem solving, if computer systems are used, is called *case based reasoning* or CBR. Instead of a set of rules, this sort of reasoning requires a large case library. The indexing system must also be of very high quality. Albeit this is a computer learning paradigm, in the case of medical expert systems (Section 19.4.1), this problem solving approach is often very useful. Here the learning may combine both inductive (several positive and negative examples) or explanation based learning (EBL), where a single example is enough to generalise a solution. For clinical decision making, iterative dichotomisers (ID) or decision tree form of inductive learning and EBL methods may be used by computers.

Such a system must find out: (i) how cases are organised in memory, (ii) how relevant cases are retrieved from memory, (iii) how previous cases can be adapted to new problems and (iv) how cases are originally acquired.

For organising cases in memory, a rich indexing system must be used.

For retrieving and matching the best case from a huge database of cases, *preference heuristics* may have to be used. They may be:

1. Goal-directed preference: preferring cases that involve the same goal as the current case.
2. Salient-feature preference: preferring cases that match the most important or maximum number of features.
3. *Specificity preference*: preferring cases that match features exactly rather than generally.
4. *Frequency preference*: preferring frequently matched cases.
5. *Recenctly matched preference*: preferring recently matched cases.
6. *Ease-of-adaptation preference*: preferring cases with features that are easily adapted to new situations.

As even the best case will not match the current situation exactly, it has to be adapted. This can be done by mapping new objects into old ones. For this, the case library may have to be augmented with a plan-modification library and a plan-repair module for future reference.

Finally, for acquiring original cases, a thorough CBPR (Section 19.3) is absolutely necessary.

19.7 Miscellaneous Use of Computer in Medicine

19.7.1 Medical transcription

If we trace the historical origin of medical transcription, we will find that it originated as a result of a requirement laid down by the medical insurance companies of the USA. These companies insist on detailed reports of services rendered by doctors and hospitals. Because of the exorbitant prices of essential healthcare services there, most people rely on the insurance agencies, and demand these detailed reports while submitting claims for reimbursement.

For each and every patient treated, preparing such extensive reports will be an extremely time- and resource-consuming exercise for the medical professionals, as well as the service providers. So, the medical and paramedical health personnel just record the essential information about each case on digital media or tapes. These dictations are then sent to professionals (medical transcriptionists), who listen, interpret and structure the dictations into full fledged reports, consisting of patient assessment, putative diagnoses, definitive diagnostic procedures undergone, therapy instilled, follow-up and so forth.

The whole process of converting healthcare dictations into comprehensive reports is known as *medical transcription (MT)*. This sort of a job requires excellent listening skills, accurate typing/word processing ability, well-written English and of course rather extensive knowledge of medical terminology.

Initially, most of these jobs were taken up by professionals from Mexico, Israel and Ireland. Unfortunately, Mexico does not have sufficient English speaking professionals, while the others countries charged exorbitantly for their services. Therefore, the focus shifted to Philippines and the Indian subcontinent. India, apart from possessing a large pool of English speakers, has an ideal time difference with the USA and an excellent telecommunications network. This is the main reason why it is becoming the new MT hub of the world.

19.7.2 Statistical analyses

Some of the statistical methods which are described in Chapter 20 (and also those which are not mentioned there) are better done with computers, especially multivariate analyses and higher order calculations. Earlier, only the simpler statistics that were calculated by pocket calculators (within a few hours) were used in research. Now, computers can solve complex problems in seconds (which previously would have taken days or months).

Interestingly, not only statistical analyses for research findings, but also day-to-day applications like CT scan or MRI (Chapter 18) are done by computers with very high computational power.

Another important area of biostatistics is the meta-analysis of various controlled cases to reach a rational conclusion for EBM (Section 19.6.1).

19.7.3 Component of diagnostic and monitoring systems

Most of the diagnostic and monitoring systems of modern day health care (e.g., the modern armamentarium in the ICU) are computer based. The details about these systems are dealt with in the respective chapters.

For a discussion on virtual reality (VR) see Chapter 3. Bioinformatics related to proteomics, genomics and drug synthesis are not discussed in this book.

This anecdote will show the utility of medical informatics and telemedicine

The Clinical Practice Library of Medicine (CPLM) was developed by Prof. Ralph R Grams and his colleagues at the University of Florida, Gainesville, under a contract with the US National Aeronautics and Space Administration (NASA). This 'Digital Doctor' is essential to the space mission crew members attempting to diagnose and treat an illness or even perform a minor surgery, while in the space. The library was informally tested several years ago, by a public health worker visiting the isolated Pacific island of Truk, (population 21000), which had no resident physician. The health worker found a group of people suffering from an illness which he could not identify. So, he plugged his portable computer into the island's telephone exchange and sent the patients' symptoms directly to Gainesville. The CPLM identified the disease as typhoid fever and recommended the treatment.

Though Telemedicine is relatively new in India, Amrita Institute of Medical Sciences and Research Centre (AIMS), Kochi, has the distinction of conducting the first telesurgery in India, which saved a Sabarimala pilgrim's life. The reader may see the relevant websites <http://www.aimshospital.org/telemed.html> and <http://www.aimshospital.org/Whatsnew/telemed.html> for further details

19.8 Further Reading

[1] Sarbadhikari, S.N. Basic medical science education must include medical informatics. *Indian J Physiol. Pharmacol.* 48(4): 395-408. 2004

[2] Sarbadhikari, S.N. The state of medical informatics in India: A roadmap for optimal organization. *J. Medical Systems* 29: 125-140. 2005.

[3] Greenes, R. A. and Shortliffe, E. H. Medical informatics: An emerging academic discipline and institutional priority. *Journal of the American Medical Association* 263: 1114–1120. 1990.

[4] Sarbadhikari, S. N. and Pal, S. K. Automated techniques for identifying depression from EEG. In *Handbook of Computational Methods in Biomaterials, Biotechnology & Biomedical Systems.* Ed. C.T. Leondes. Kluwer Academic Publishers. 2002 (In Press)

[5] Sarbadhikari, S. N. Will the aging computer technology comfort the aging humans?. *Speculations in Science & Technology* (Special Issue on the Computers in the Fight Against Aging) 19(2): 131–136. 1996.

[6] Sarbadhikari, S. N. Medical informatics — Are the doctors ready?. *J. Indian Med. Assoc.* 93: 165–166. 1995.

[7] Sarbadhikari, S. N. Let your dreams come true through virtual reality (current topic). *J. Indian Med. Assoc.* 95: 190. 1997.

[8] American National Standards Institute (ANSI) Healthcare Informatics Standards board (HISB). *Inventory of Clinical Information Standards.* New York. 1998 [http://web.ansi.org/rooms/room_41/public/docs.html]

[9] CTICM. *Guide to Software & Resources for Computer Based Learning in Medicine, Dentistry, Pharmacy & Veterinary Sciences.* 3rd ed. 1996, www.ets.bris.ac.uk/cticm/calguide.htm; www.ets.bris.ac.uk/calscribe

[10] *Learning Technology Dissemination Initiative (LTDI) Information Directory.* 8th ed., 1996, www.icbl.hw.ac.uk/ltdi/handbook/hbtot.htm

[11] MIDRIB (C.Lombardo@sghms.ac.uk; jules.cook@bristol.ac.uk)

[12] Computer-based Patient Record Institute (CPRI). *Position Paper Computer-based Patient Record Standards.* Chicago, IL. 1994.

[13] Moshman Associates Inc. Current activities of selected health care informatics standards organizations. Office of Science and Data Division, Agency for Health Care Policy and Research, Bethesda, MD. 1994.

[14] Shortliffe, E.H., Fagan, L.M., Wiederhold, G. and Perreault, L.E. *Medical Informatics: Computer Applications in Health Care and Biomedicine (Health Informatics).* 2nd ed. Springer Verlag. 2000.

[15] Bemmel, J., Van Bemmel, J. and Musen, M.A., eds. *Handbook of Medical Informatics.* Springer Verlag. 1997.

[16] Berner, E.S. and Ball, M.J., eds. *Clinical Decision Support Systems: Theory and Practice.* Springer Verlag. 1998.

[17] Rada, R. *Information Systems for Health Care Enterprises.* 2nd ed. HIPAA IT LLC. 2003.

[18] Eder, L.B., ed. *Managing Healthcare Information Systems with Web-Enabled Technologies.* Idea. 2000.

[19] www.mihandbook.stanford.edu/handbook/home.htm or www.mieur.nl/mihandbook/r_3_3/handbook/home.htm

[20] http://www.syedzia.net/blogs/indiamedinfo/archives/cat_medical_informatics.html

[21] www.cpmc.columbia.edu/edu/textbook/
[22] http://smi-web.stanford.edu/textbook/
[23] www2.utsouthwestern.edu/cfdocs/library/ links/toolboxes/medicalInformatics.cfm
[24] www.keukahealth.com/medical_professional/ medical_informatics.htm
[25] www.ohsu.edu/bicc-informatics/hershbook.shtml
[26] www.ehto.org/ikb/urls/cneh/infmedwo.html
[27] www.amia.org
[28] www.hl7.org (HL7)
[29] http://aspe.os.dhhs.gov/admnsimp/ (HIPAA)
[30] http://medical.nema.org/dicom.html (DICOM)
[31] www.iamindia.org
[32] www.who.int/whosis/icd10/ (ICD-10)

19.9 Questions

1. Descriptive Questions

1. How can computers help in reaching a clinical diagnosis?
2. Explain the advantages of EBM with suitable examples.

2. Write Short Notes on

1. Clinical or diagnostic decision support systems
2. Computer based patient records
3. Health information systems
4. Medical transcription
5. Computer assisted/aided learning and instructions

3. Multiple Choice Questions

Choose the single best response.

1. Medical informatics includes all except
   (A) HIS
   (B) CBPR
   (C) Biomedical instrumentation
   (D) CDSS/DDSS

2. Connectionist expert systems are useful for medical diagnosis because:
   (A) natural and linguistic inputs are plentiful in medical jargon
   (B) inferencing and rule generation may help in new knowledge formation of the advancing medical sciences
   (C) often medical history and examination give only subjective and qualitative diagnostic criteria.
   (D) all

3. CBPR is not utilised fully if it is
   (A) not connected to other health care centres
   (B) in separate modules
   (C) not using latest versions of software and hardware
   (D) none of the above.

4. CAI may not be useful if it
   (A) does not use the latest versions of software and hardware
   (B) is done without proper supplementation of textbooks
   (C) is done under the supervision of teachers
   (D) none of the above.

5. Medical transcription does not require
   (A) good medical knowledge
   (B) good knowledge of medical terminology
   (C) excellent listening power
   (D) good word processing capacity

Answers

1. C; 2. D; 3. A; 4. B; 5. A

Section 8

Biostatistics

Biostatistics deals with the collection, classification and tabulation of numerical biological/biomedical facts as the basis for description, explanation and comparison of a phenomenon.

Statistics have been maligned by all and sundry. It is said that there are three types of lies: lies, damned lies and statistics. Another definition of statistics is that, like a swimsuit, it reveals the bare facts but conceals the vitals. Why does statistics have such a bad name? It is because statistics has a way of presenting data that makes it useful or useless (to the presenter, as well as the user). The single chapter (20) in this section deals with various basic statistical procedures for analysing, comparing and interpreting biomedical data.

20

Basic Statistical Methods for Analysing Biomedical Data

20.1 Introduction

The representative value from a sample is called a *statistic*, while that from a population is called a *parameter*. The approximate (reference) ranges of values in humans, for many measurements are not yet standardised. A single standard nomenclature using SI units is still not universally acceptable. The accuracy of methods for laboratory investigations varies, due to error in samples and also methodical error. Moreover, values vary from one normal individual to another because of biologic variation. Some variables like age, sex, time of day, time since last meal, can be accounted for. However, many others variables cannot be standardised. The facts collected in a study are called *(raw) data*. In this chapter, we shall deal with only numerical data. Processed data (in an understandable and usable form) is called *information*. When we collect any data, it has to be put in the form of a table or tabulated.

20.2 Representation of Data

Data may be tabulated simply (alphabetically or chronologically) or as a frequency distribution table having proper groups (class intervals) and the number of time each event occurs (frequency). Tables are neither very attractive nor impressive. So, it is better to represent the same data in figures like bar charts (simple, multiple, component and pictogram), histogram (e.g., runs per over in a one-day cricket match), frequency polygon, line diagrams and pie charts. Scatter diagrams or scatter-plots show the relation between two variables.

20.2.1 Central Values and Dispersion

In order to get an idea of the *central value* of a collection of data, (arithmetic) mean $M = (\sum x/n)$, median (*Mdn*) (middle value of the observations sorted in ascending or descending order) and mode (M_0) (most frequently occurring value) are usually calculated. The Mean *M* may be denoted by $\bar{x}$. Often $M_0 = 3Mdn - 2M$.

For measuring *dispersion* (variation from the central value), range, mean or average deviation and standard deviation are used. Mean deviation (MD) is $(\sum(x-M) / n)$.

The standard deviation (SD) (also known as the root mean square or RMS deviation) of the sample (*s*) around the (arithmetic) mean can be measured by

$$s = \sqrt{\frac{\sum(M-x)^2}{n-1}}$$

where *M* is the arithmetic mean and *x* is the individual value, *n* being the sample size, for a sample size of $n \leq 30$. If $n > 30$, the denominator can be *n*. SD of the mean of the whole population is denoted by s. For a truly representative sample, *s* and s will be comparable.

A frequency distribution (FD) curve may be constructed from individual values in a population by plotting the frequency with which any particular value occurs in the series against the values. For a homogenous group (large population with narrow class intervals), the FD curve follows a *normal* distribution. Albeit there are infinite number of normal curves according to the mean and SD, there is only one *standardised* normal curve based on an infinitely large number of observations. This curve (Fig. 20.1) is smooth, perfectly symmetrical, bell shaped or Gaussian, with mean = mode = median = 0, SD = 1 and area under curve (AUC) = 1. Usually $M \pm 1\sigma$ contains 68.27 – 68.3% of values in distribution, $M \pm 2\sigma$ covers 95.4 – 95.45% of values in distribution ($M \pm 1.96\sigma$ contains 95.00% of values) and $M \pm 3\sigma$ contains 99.7 – 99.73% of values in distribution. The limits on either side of the mean are called *confidence intervals* (Fig.20.2). For instance, the probability of a (random) reading falling outside 95% confidence limit ($M \pm 1.96\sigma$) is 1 in 20 ($p = 0.05$).

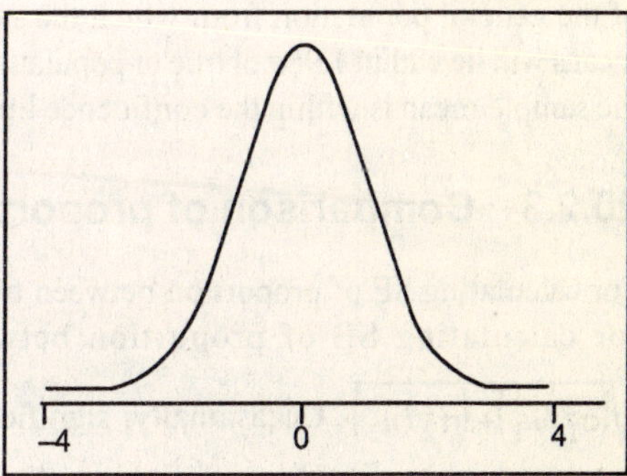

Fig. 20.1 Standardised normal curve

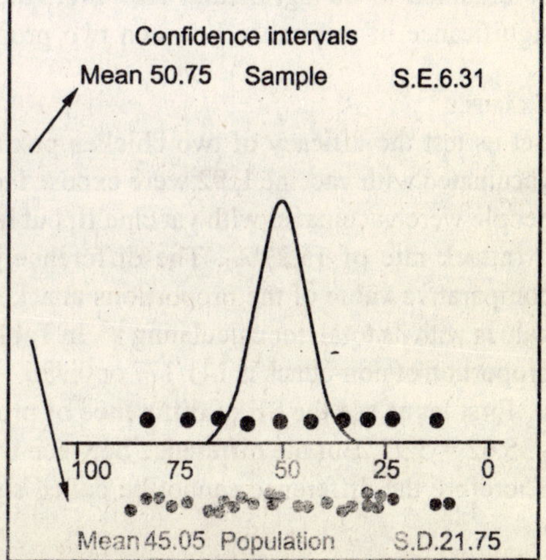

Fig.20.2 Confidence intervals

This means that there are 5% chances that the value is normal and 95% chances (19 out of 20) that the value is abnormal. In the FD curve, $(x - M)$ distance, in terms of SD is called the *relative deviate* or *standard normal variate* $Z = (x - M)/\sigma$. Z is a standardised variable and similar to x follows a normal distribution.

20.2.2 Sampling

Instead of studying the whole population or universe, it is more easy and economical to study a sample (which of course should be representative). To reduce the effects of confounding factors during sampling, it is advisable to randomise the readings. The sampling method may be simple random, systematic random (e.g., picking up every n^{th} unit at regular intervals) or stratified random (e.g., according to age groups). Variation between samples is due to sampling error – commonly because of the small size of the sample and also due to recall bias. Conventionally, Greek letters denote population mean (μ), SD (σ) and Roman letters indicate sample mean (M), SD (s).

Non-sampling errors usually result due to instrumental faults and observer inadequacy.

SD of the sample means is a measure of sample error, and is known as the *standard error of the mean* or SEM ($\sigma/\sqrt{n}$) that strictly indicates the reliability of the sample mean as a representative of the true mean of the general population from which the sample was drawn. In a normal distribution, 95% of sample means will lie within 1.96σ of true or population mean $[\mu \pm 2(\sigma/\sqrt{n})]$. So, SEM is a measure to judge whether the sample mean is within the confidence limits or not.

20.2.3 Comparison of proportions

For calculating SE of proportion between two proportions p and q, SE(P) is calculated as $\sqrt{(pq/n)}$, and for calculating SE of proportion between two means (σ_1 and σ_2), SE(d) is denoted by $\sqrt{(\sigma_1^2/n_1)+(\sigma_2^2/n_2)}$. Occasionally, significance of difference between two proportions or ratios has to be measured by the formula $\sqrt{(p_1q_1/n_1 + p_2q_2/n_2)}$. Here, only if the observed difference between two groups (e.g., p_1 and p_2) is greater than twice the SE of the difference of proportions, can the difference be assumed to be significant. However, the Chi-square (χ^2) test is a better method for testing the significance of difference between two proportions.

Example

Let us test the efficacy of two chicken pox (*Varicella zoster*) vaccines I and II. Of the 2500 people vaccinated with vaccine I, 92 were exposed to the disease and 23 contracted it (25% attack rate). 2200 people were vaccinated with vaccine II, out of which 85 were exposed to the disease and 13 contracted it (attack rate of 15.29%). The difference between the attack rates is 9.71%. Table 20.1 gives a comparative value of the proportions attacked (p) and not attacked (q). Table 20.2 presents the same values with its total, for calculating χ^2. In Table 20.2, the proportion of attack is 36/177 or 0.203, and the proportion of non-attack is 141/177 or 0.796.

First let us see the SE of difference of proportions. $\sqrt{(p_1q_1/n_1 + p_2q_2/n_2)}$ will be $\sqrt{(20.38 + 15.24)} = \sqrt{35.62} = 5.97$. But the difference between the attack rates is 9.71%, which is less than twice of 5.97. Therefore the difference cannot be called significant.

Table 20.1 Proportion of attacks

Vaccine	Attacked (p)	Not attacked (q)
I	25%	75%
II	15.29%	84.71%

Table 20.2 Attack rates with 2 vaccines

Vaccine	Attacked	Not attacked	Total	Attack rate
I	23	69	92	25%
II	13	72	85	15.29%
Total	36	141	177	–

20.2.4 Chi-square test (Non-parametric)

In the χ^2 test initially a *null hypothesis* that there is no difference between the effects of two treatments has to be set. Then, χ^2 test is applied as $\chi^2 = \sum (O - E)^2 / E$, where O is the observed frequency and E is the expected frequency or $f_e = ff_c / n$, (calculated by pooling the results from two treatments, given in the rows and columns). The d*egree of freedom (DOF or df)* of a statistic gives the number of cases or classes in a sample that is free to vary during computation, without affecting sample size and is used to estimate the corresponding population parameters. It is found by multiplying ($\chi - 1$) and ($r - 1$). So, in a 2 × 2 contingency table, *df* is 1. After calculating the χ^2, its probability for that df is seen from the standard table. If the observed χ^2 value is higher, significance is assumed.

Now let us take the values from Table 20.2 and arrange it in Table 20.3.

Table 20.3 Table for Chi-square test

Vaccine	Attacked	Not attacked
I	$O = 23$	$O = 69$
	$E = 92 \times 0.203 = 18.676$	$E = 92 \times 0.796 = 73.232$
	$O - E = 4.324$	$O - E = -4.232$
II	$O = 13$	$O = 72$
	$E = 85 \times 0.203 = 17.255$	$E = 85 \times 0.796 = 67.66$
	$O - E = -4.255$	$O - E = 4.34$

Applying the null hypothesis and $\chi^2 = \sum(O - E)^2 / E = 18.69/18.676 + 17.91/73.232 + 18.11/17.255 + 18.84/67.66 = 1.0 + 0.24 + 1.05 + 0.28 = 2.57$. However, the value of χ^2 with df 1 and at $p = 0.05$ is 3.84. Therefore the null hypothesis stands and there is no statistically significant difference between the two vaccines I and II for preventing chicken pox.

20.2.5 Student's *t*-test (parametric)

For evaluating significance of difference between two means, in biomedical research, measurements are often made on a group of subjects (human or animal) undergoing some treatment. These values are compared with similar measurements on a control group with similar characteristics (*ceteris paribus*

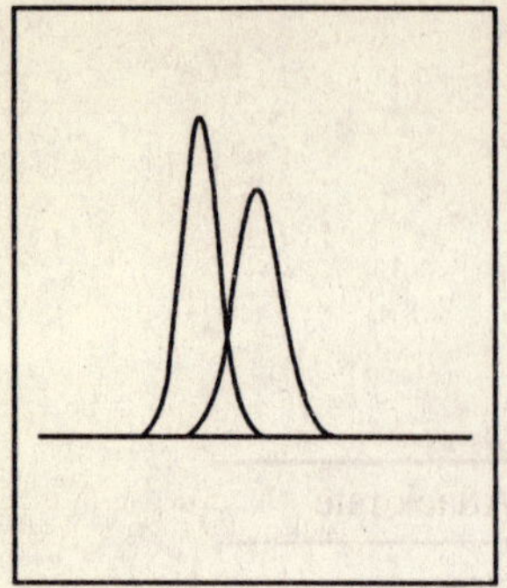

Fig. 20.3

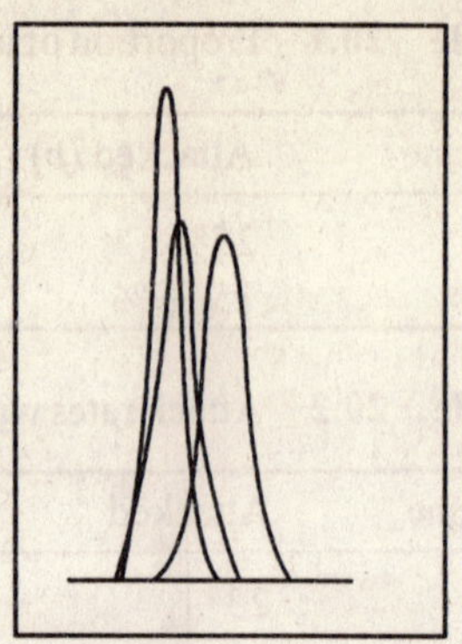

Fig. 20.4

or other things remaining equal) but without treatment. If a particular mean value in the treated group is different from the corresponding mean in the control group, one has to ascertain whether the difference is due to treatment or chance variation. The probability that the difference between two groups (e.g., with and without treatment) represents chance variations can be estimated in most cases with a student's *t*-test (Fig.20.3). The value *t* is the ratio of difference of means of 2 series M_1 and M_2 to the uncertainty in these means.

For n_1 values in *Series 1* and n_2 values in *Series 2*,

$$t = \frac{M_1 - M_2}{\sqrt{\dfrac{(n_1+n_2)[(n_1-1)s_1^2+(n_2-1)s_2^2]}{n_1 n_2 (n_1+n_2-2)}}}$$

when $n_1 = n_2$, $t = (M_1 - M_2)/\sqrt{((SEM_1)^2 + (SEM_2)^2)}$

A *2-tail t-test* measures the significance in both directions (positive and negative), while a *1-tail t-test* measures the significance in only one direction (either positive or negative). The critical *t* value's probability of chance occurrence is determined by a (level of significance). For a 2-tail *t*-test, an area equivalent to $\alpha/2$ lies beyond the critical *t* value in each tail of distribution, for that *df*. On the other hand, an area α is located beyond the critical *t* value in a single tail.

Higher the value of *t*, less is the probability that the difference represents chance variation. The probability also reduces with increase in the number of sample size (lower sampling error). The *p* values for *t* at different values of *n* can be found in standard tables of statistics (e.g., Fisher and Yates, 1974). A *p* value of <0.001 implies that the chances of the difference being due to random variation are less than 1 in 1000. Statistically significant difference is usually taken at *p* values < 0.05 (however if it is <0.01 or <0.001, it is definitely more significant) – due to the operation of some factor, rather than chance.

Example

In an experimental group of *Rattus norvegicus var albinus* (albino variety of Norway rat), with 10 controls and 10 treated subjects, an anti-obesity drug was tried for 15 days @ 10mg/kg/d. on the treated group, without administering the drug to the control group. Both the groups were given food *ad libitum* (as much as they wanted). The findings are tabulated below (Table 20.4).

Table 20.4 Body weight of rats with and without treatment

	Body weight of the rats before treatment (g)									
Control	105	95	85	102	108	103	95	87	102	90
Treated	103	97	87	107	110	105	97	90	105	92
	Body weight of the rats after 15 days' treatment (g)									
Control	110	101	90	110	112	107	98	91	106	95
Treated	101	95	85	103	108	102	95	88	101	90
	Gain in weight by the control group									
D_g	+5	+6	+5	+8	+4	+4	+3	+4	+4	+5
D_g^2	25	36	25	64	16	16	9	16	16	25
	Loss in weight by the treated group									
D_l	−2	−2	−2	−4	−2	−3	−2	−2	−4	−2
D_l^2	4	4	4	16	4	9	4	4	16	4

$\sum D_g = 48$, Mean $D_g = 4.8$, $\sum D_g^2 = 248$, $SD_g = \sqrt{(248 - 48^2/10)} / df(9) = 1.4$,

$SE_{Dg} = 1.4 / \sqrt{10} = 0.44$, $t =$ Mean $D_g / SE_{Dg} = 4.8 / 0.44 = 10.91$.

$\sum D_l = -25$, Mean $D_l = -2.5$, $\sum D_l^2 = 69$, $SD_l = \sqrt{(69 - 25^2/10)} / df(9) = 0.85$,

$SE_{Dl} = 0.85 / \sqrt{10} = 0.27$, $t =$ Mean $D_l / SE_{Dl} = 2.5 / 0.27 = 9.26$.

Now, the tabulated t for df 9 is 4.78. Therefore, the inference will be:

(i) Mean gain in weight by control animals (4.8 g) has $t = 10.91$, while tabulated value is 4.78 at $p = 0.001$. In other words, weight gain in control animals is significant after 15 days (without any treatment) at $p = 0.001$.

(ii) Mean loss in weight in treated animals (2.5g) has $t = 9.26$, while tabulated value is 4.78 at $p = 0.001$. That means that loss of weight in treated animals is significant, after 15 days of treatment with the drug, at $p = 0.001$.

PAIRED *t*-TEST

This test is usually made between similar individuals or items. Often the same individual serves as a pair when measurements are made at different times of the experiment. This is called *self-pairing*. By null hypothesis, it is assumed that the difference between the pairs is zero. If the experimental data shows any difference, the statistical testing of significance is done by $t = \overline{D} / (S_D/\sqrt{n})$, where the denominator is SE of the difference (SE_D), D is the difference between individuals of the pair, n is the number of pairs, $\overline{D} = \sum D/n$ or mean of deviations, $S_D = \sqrt{\left(\sum D^2 - \left(\sum D\right)^2 / n\right) / (n-1)}$ and df is (n–1).

However, pooled σ_d^2 has $2(n-1)$ or $2n-2$ df. So the tabulated t value relating to independent samples will be of a lower magnitude than the tabulated t of the paired data. The advantage of paired t is that it does not require σ_1 to be equal to σ_2.

20.2.6 Analysis of variance (ANOVA)

For more than two groups, to avoid systematic error, *an*alysis *o*f *va*riance (ANOVA) is the test of choice. Variance or mean square (s^2 or MS) is the sum of squared difference between the mean and value $\sum(M-x)^2$. For computing σ^2, sum of squares (*SS* or *A*) is required. The value being a squared one, $(M-x)^2$ and $(x-M)^2$ are identical.

$$s^2 = \frac{\sum(x-M)^2}{n} = \frac{\sum x^2}{n} = \frac{SS}{n} = D \text{ (For } n<30\text{, denominator should be } n-1\text{)}$$

Here n denotes the total number of observations i.e., $n_1+n_2+...+n_m$ or the sum of n_1 observations from group 1 through n_m observations from group m. We shall assume that $n_1= n_2= ...=n_m$. Otherwise, if the sample size in the various groups is unequal, the calculation becomes much more complicated and is beyond the scope of this book. The ratio of variance to mean (s^2/M) is called the *coefficient of dispersion* (CD). For population variance, s^2 is used.

The analysis is directed to (a) variations between treatments – usually shown in the columns (*e.g.*, *m* groups) of the table and (b) variations in individual readings of a particular item or *replicates* (similar individual or items receiving identical treatment) shown in the rows of the table. *F*-statistic is the ratio of variances and can be calculated and compared with standard *F*-tables (e.g., Fisher and Yates, 1974).

Steps for calculation

(i) Sum up squared deviations of all observations for *n* observations and subtract the mean:

$SS - SS/n$ or $(A-D)$.

(ii) Sum up squared deviations due to treatment:

$(SS_1/n_1+SS_2/n_2+...+SS_m/n_m-SS/n)$ or $(B-D)$.

(iii) Sum up squared deviations due to random differences in items of treatment:

$SS - (SS_1/n_1+SS_2/n_2+...+SS_m/n_m)$ or $(A - B)$

(A) Variation due to treatment only: $K_c = (SS_1/n_1+SS_2/n_2+...+SS_m/n_m -SS/n)$ / df
Here *df* = (number of treatments or number of columns minus one). In other words, K_c is nothing but the mean square (MS) of columns.

(B) Variation due to items within treatment (residual variation):

$K_r = (SS - (SS_1/n_1+SS_2/n_2+...+SS_m/n_m))$ / df

Here *df* = treatments × (number of items within treatments or number of rows minus one). That means K_r is the MS of column ′ (rows – 1).

$F = K_c / K_r$

Example

In an experimental group of subjects, with 5 controls (C) and 5 each in three experimental groups: D (stressed and depressed), EO (exercised only) and ES (exercised and stressed), the REM (rapid eye movement or dream phase) sleep percentage compared to the total sleep was measured. The findings are tabulated below (Table 20.5).

Table 20.5 Percentage of REM sleep in 4 groups (C, D, EO and ES) of subjects (Mean=12.2)

	C			D			EO			ES		
		M–x	$(M–x)^2$		*M–x*	$(M–x)^2$		*M–x*	$(M–x)^2$		*M–x*	$M–x)^2$
1	12.3	0.1	0.01	15.7	3.5	12.25	11.9	0.3	0.09	12.0	0.2	0.04
2	12.1	0.1	0.01	15.8	3.6	12.96	11.8	0.4	0.16	11.9	0.3	0.09
3	12.7	0.5	0.25	16.1	3.9	15.21	12.0	0.2	0.04	12.2	0	0
4	11.9	0.3	0.09	16.2	4	16	11.7	0.5	0.25	11.8	0.4	0.16
5	12.0	0.2	0.04	15.9	3.7	13.69	11.9	0.3	0.09	12.2	0	0
Sum	–	1.2	0.4	–	18.7	70.11	–	1.7	0.63	–	0.9	0.29

In this example,

$$SS = A = 0.4+70.11+0.63+0.29 = 71.43,$$

$$B = (1.2^2+18.7^2+1.7^2+0.9^2)/5 = 70.97,$$

$$D = (\sum x)^2/(n-1) = (1.2+18.7+1.7+0.9)^2/(20-1) = 22.5/19 = 1.18.$$

Now see Tablc 20.6 for calculation of the *F* value.

Table 20.6 Table for ANOVA

Source of variation	*SS*	df	MS
Between columns (groups)	$B–D$= 69.79	$c–1$= 3	$K_c = (B–D)/(c–1)$= 23.26
Residual (between replicates)	$A–B$= 0.46	$c(r–1)$= 16	$K_r = (A–B)/c(r–1)$ = 0.03

Therefore, the calculated *F* value is 23.26/0.03 = 775.33. The *F* values tabulated for *df*(3,16) are 3.2, 5.3 and 9.0 respectively for *p* values 0.05, 0.01 and 0.001. As the obtained value is far higher than the tabulated values, the differences are statistically significant even at $p = 0.001$ (1 in 1000).

Conventionally, to show significance levels of the columns, * is put when the calculated $F < 0.05p$, ** is put when the calculated $F < 0.01p$ and *** is put when the calculated $F < 0.001\ p$.

For multivariate analysis, now MANOVA or multivariate analysis of variance is used.

Values missing in a table for ANOVA: This can be calculated by

$$x = (rR + cC - T) / (r-1)(c-1)$$

where r = number of rows, c = number of columns, R = total of the row where a value is missing, C = total of the column where a value is missing and T = the grand total of the values, taking into account the ± signs.

For testing the significance of difference between two or more proportions, χ^2 test is useful. It is a non-parametric (distribution free) test like Spearman's ρ, Kendall's τ and Wilcoxon's rank test.

20.3 Association between parameters

20.3.1 Correlation coefficient

For estimating the significance of association between two variable parameters, coefficient of correlation (r) testing is appropriate. For instance, for two variables x and y in n individuals, each has one reading of x and y,

$$r = \frac{\sum(x - M_x)(y - M_y)}{\sqrt{\sum(x - M_x)^2 \sum(y - M_y)^2}}$$

If r tends to +1, a strong positive association between x and y is reflected i.e., if x increases, y also proportionally increases. On the other hand, if r tends to −1, a strong negative association between x and y is reflected i.e., if x increases, y proportionally decreases. In some cases, there may be a pseudo correlation or spurious association (e.g., rainfall and admission in colleges during monsoon).

20.3.2 Regression

To purvey means of estimating the value of one variable from the value of another, regression analysis is performed. In the case of an individual, to know the value of one variable, when the value of the other variable is known, regression coefficient of one is measured with respect to the other. Conventionally, the independent variable is denoted by x and the dependent variable is represented by y. Suppose M_y = mean of $y_1, y_2, \ldots, y_n$ and M_x = mean of $x_1, x_2, \ldots, x_n$. Then, $y = M_y + b_y(x - M_x)$,

where b_y (regression coefficient of y upon x) $= \dfrac{\sum(x - M_x)(y - M_y)}{\sum(x - M_x)^2}$.

Similarly, for regression coefficient of x upon y, $b_x = \dfrac{\sum(x - M_x)(y - M_y)}{\sum(y - M_y)^2}$.

Therefore, $x = M_x + b_x(y - M_y)$

For calculating multiple regression, programs as those given in Appendix D.10 may be used.

20.3.3 Nomogram

Nomogram is a graphical method for predicting values of a variable, in an individual, from the values of some other variables bearing a correlation with the predicted variable. It comprises a graphical chart, carrying a number of fixed, parallel or non-parallel, straight (rectilinear) or curvilinear scales, each for the values of one of the variables, and bearing either linear or logarithmic graduations. Computed in accordance with the mathematical equations describing the relation between these variables, scales are graduated. The scales are positioned in such a way as to make the straight line joining the two points A and C on the edges of the first and third scales respectively, intersect the second scale at point B. This gives the predicted value of the corresponding variable in an individual having the values of the other variables as A and C. An example of a nomogram is that used for calculating body surface area given the height and weight of the individual. The scales for children and adults are, however, different. Similarly, a very useful curve is the Siggaard–Andersen nomogram (where the P_{CO2} or partial pressure of carbon dioxide gas, in our arterial blood is plotted on a logarithmic scale in the X-axis and pH is plotted linearly in the Y-axis). The horizontal line through P_{CO2} at 40 mm of Hg divides areas into effective degrees of hypo- or hyper-ventilation to maintain the acid- base balance of our body.

Some excellent examples of nomogram are available on the following sites <http://www.bioscience.org/atlases/clinical/nomogram/nomoadul.htm>, <http://www.hc-sc.gc.ca/hpfb-dgpsa/onpp-bppn/cg_quick_reference_e.html> and <http://www.cebm.net/likelihood_ratios.asp>

20.3.4 Factor analysis and cluster analysis

Some useful tests are available for classification, especially in psychology and education. One such test – factor analysis (FA) starts with a set of observations obtained from a given sample by means of some *a priori* measures. Spearman was one of the pioneers of this method. A set of observations are analysed from their intercorrelations to determine whether the variations represented can be sufficiently accounted for by a number of basic categories smaller than the initial categories considered while starting the experiment. Therefore, data obtained with a large number of *a priori* measures may be explained in terms of a smaller number of reference variables.

Cluster analysis (CA), a simple form of correlational analysis, another useful technique, may also be used for better understanding of the purposes and nature of FA. For 'pure' variables (each appearing on only one common factor), CA and FA give similar results and CA is easier to perform. While in CA, each variable as a unit is placed in a cluster, in FA, different portions of the variance of a variable may be assigned to different factors. These sorts of analyses are done on a matrix of intercorrelations. However, detailed discussions are beyond the scope of this text.

20.4 Further Reading

[1] Dawson-Saunders, B. and Trapp, R. G. *Basic and Clinical Biostatistics* 2nd ed. Appleton & Lange. 1994.

[2] Knapp, R. G. and Miller, M. C. III, *Clinical Epidemiology and Biostatistics.* Baltimore, MD: Williams & Wilkins. 1992.

[3] Puri, B. K. *Statistics for Health Sciences* (using SPSS). WB Saunders. 1996.

[4] Anthony, D. *Understanding Advanced Statistics.* Edinburgh: Churchill Livingstone. 1999.

[5] Lwanga, S. K. and Tye, C-Y. *Teaching Health Statistics.* WHO, 1986. Authorized Reprint by CBS Publishers, New Delhi, 1994.

[6] Fisher, R. A. and Yates, F. *Statistical Tables for Biological, Agricultural and Medical Research,* 6th ed. Longman Group Ltd. 1974.

20.5 Questions

1. Descriptive Questions

1. What are the measures of dispersion?
2. Describe a normal curve and discuss its significance.

3. Describe a common non-parametric test with its calculations.
4. Give a detailed description of a Student's *t*-test.
5. What is ANOVA ? How do you calculate the *F* value and draw any inference from it?

2. Short Notes

1. Standard deviation
2. Standard error of the mean
3. Correlation coefficient
4. Regression analysis
5. Paired *t*-test
6. 2-tailed *t*-test
7. χ^2-test
8. $p=0.05$
9. Level of significance
10. Nomogram.

3. Multiple Choice Questions

1. A suitable chart to depict clearly the decadal total census is
   (A) Histogram
   (B) line diagram
   (C) simple bar chart
   (D) frequency polygon

2. The average birth weights in a hospital are best demonstrated statistically by
   (A) bar chart
   (B) histogram
   (C) pie chart
   (D) frequency polygon.

3. A scatter diagram shows
   (A) trends of events over time
   (B) relationship between 2 variables
   (C) proportion of an event
   (D) positive relationship

4. Determination of which statistical parameter requires quantities to be arranged in ascending or descending order?
   (A) mean
   (B) median
   (C) mode
   (D) SD

5. The most frequently used measure of dispersion is
   (A) mean (B) median (C) standard deviation (D) range

6. Which is true of SEM?
   (A) It is the SD of the mean of the sample. (B) *pq/n* (C) It can be used to differentiate the efficacy of 2 vaccines (D) none of the above

7. In a normal curve
(A) mean = SD (B) mean = median (C) mean = 2SD (D) mean = variance

8. Confidence limits can be calculated using
(A) mean and range (B) mean and SE (C) median and range (D) median and SE

9. Which statement about standard deviation of a sample is incorrect?
(A) It is the geometric mean of the deviations
(B) It is the root mean square deviation.
(C) It is denoted by *s*.
(D) While calculating, the denominator should be $(n-1)$ for a sample size > 30.

10. 95% confidence limits for a normal curve is a value around the mean
(A) 1.96σ (B) 2σ (C) 3σ (D) 4σ

11. 95% of values in a normal curve lie between
(A) 1σ (B) 2σ (C) 3σ (D) 4σ

12. The term $\sqrt{(pq/n)}$ indicates
(A) SEM (B) SE of difference of means
(C) SE of proportions (D) SE of difference of proportions.

13. If the mean is 230 and SE is 10, 95% confidence limits would be
(A) 210–250 (B) 220–240 (C) 225–235 (D) 230 ± 15

14. Randomisation is useful to eliminate
(A) observer bias (B) confounding factors (C) subject bias (D) sampling bias

15. The test of significance used while investigating discrete variable is
(A) χ^2 - test (B) *t*- test (C) SE of difference between 2 means (D) none of the above

16. All of the following statistical tests are non-parametric except
(A) correlation (B) rank correlation (C) Wilcoxon rank sum test (D) student's *t*-test

17. The mean BP of a group of persons was determined and after an interventional trial, the mean BP was estimated again. The test to be applied to determine the significance of intervention is
(A) χ^2-test (B) paired *t*-test (C) correlation coefficient (D) mean deviation

18. A '2 × 2' contingency table construction is applicable for calculation of all except
(A) relative risk (B) validity of a screening test (C) odds ratio (D) regression coefficient

19. If there are more than two experimental groups, which statement is incorrect?
(A) Variance cannot be analysed. (B) There is chance of systematic error.
(C) *t*-test may not be helpful. (D) χ^2-test cannot be done.

20. Which of the following does not correlate association between parameters?
(A) nomogram (B) regression analysis (C) *t*-test (D) factor analysis.

Answers

1. A; 2. A; 3. B; 4. B; 5. C; 6. A; 7. B; 8. B; 9. D; 10. A; 11. B; 12. C; 13. A; 14. B; 15. A; 16. D; 17. B; 18. D; 19. A; 20. C.

Section 9

Safety, Planning and Ethical issues

While dealing with biomedical equipment, whether electromechanical, electromagnetic or radiating ones, some amount of precautionary measures are essential for the manufacturer, users (doctors/surgeons, nurses, technicians), as well as, the end recipients or patients. A 'stitch in time saves nine' is an appropriate adage like 'an ounce of prevention is worth a pound of cure' regarding biosafety measures.

Setting up a biomedical engineering or medical science and technology unit in any healthcare delivery system requires extensive planning in accordance with the specific needs of the system and affordability.

It is also essential to keep the ethical and social concerns in mind while setting up a clinical or research facility. After all human beings are now considered not just psychosomatic, but rather psycho-bio-social organisms. Therefore planning, in a broader sense, also has to include socioethical issues.

This section briefly deals with these aspects.

21

Safety Measures for Biomedical Equipment

21.1 Introduction

Electromedical (electronic, electromagnetic and electromechanical) equipment itself is basically harmless, but because it makes a conductive contact with the patient, it can be the key link in a life-threatening current path

Radiating equipment also involve a certain amount of risk. Radiation, above a certain amount can harm our bodies. This harm takes place at the cellular level. Radiation causes ionisations in the molecules of our body's cells. Ionisation is the removal of electrons from the atoms, which then go on to form ions (i.e., charged atoms). It is the ions that cause the cellular damage. At low doses (up to 1 Sv), the cells can repair the damage rapidly but at higher doses, the cells may not be able to keep pace with the required repair. This causes the cells either to die or be changed permanently. If the cell dies, the body quickly replaces it, leaving no ill effects. However, the cells that have permanently changed produce abnormal cells and under the right conditions, these cells can become cancerous.

Microwaves (either from the oven or from mobile/cell phones) are also harmful to our body. Although there has been some research findings inferring that the prolonged and massive use of cell phones causes damage to the protective BBB (blood brain barrier), more work is definitely required to establish the safety parameters and limits for microwaves. Therefore it will not be discussed in this chapter.

Safety measures should be carefully implemented for the manufacturers, persons operating such equipment, as well as for the end users or patients.

21.2 Electromedical Equipment

21.2.1 Physiological effects of electric shock

Electric current passing through the human body has primarily three effects: (i) injury to tissues, (ii) uncontrollable muscle contraction or unconsciousness, and, (iii) fibrillation (uncontrolled increase of beating rate) of the heart. The common effects are tabulated in Table 21.1.

Table 21.1 Effect of a 50 Hz electric current on an average human body (contact of 1 second duration)

Current Intensity	Effect
1 mA	Threshold of perception
5 mA	Accepted as maximumharmless current intensity
6–50 mA	'Let go' current before sustained muscular contraction
50–100 mA	Pain, possible fainting, although heart and respiratory functions continue
100– 200 mA (*most lethal range*)	Ventricular fibrillation (VF) starts but respiratory centres remain intact
200–350 mA	VF may or may not occur
At 6 A or around	Sustained myocardial contraction (cardiac standstill), may be followed by normal heart rhythm (so the prognosis is better than in the 100–200 mA range);Temporary respiratory paralysis
Very much above 6 A	Burns are possible due to heating effects of the current

The voltage required to obtain these current levels depends on the skin type, contact area and whether or not the skin is moist. If a person has to touch an electrical instrument while standing barefoot on wet earth, there should not be more than 5 mA current flowing through the person's body. Electric shocks may be either *macroshock* (due to contacts applied to the exterior of the body (large value electrical current) or *microshock* (when a small value leakage current passes directly to the heart).

During any procedure requiring electromedical equipment, the current passing internally through a patient is usually kept below 10 mA. Non-therapeutic contact voltages are commonly less than 5 mV.

21.2.2 Preventive measures to reduce shock hazards

During manufacture of electromedical equipment, it is expected that the makers will follow good manufacturing practice (GMP) so that the equipment are not faulty or hazardous at the time of delivery and installation.

The most important preventive measure which can be taken to avoid shocks from hospital equipment is to arrange for careful grounding. Using a line isolated transformer will enable line isolation. Unfortunately many hospitals in our country do not adhere to these time-tested measures.

The other guidelines for maintaining a safe electrical environment for the hospital staff and patients are:

1. Only use three-wire power cables
2. Supply power to all electrical equipment within 5 m of the patient from a single bank of power points that have grounded terminals tied together by a 12-gauge wire

3. Do not use any other electrical outlets in the vicinity of the patient
4. Supply power points from a power line isolation transformer
5. Connect the conductive surfaces of all non-electrical devices/accessories to a common ground terminal
6. Check power cords from the plug to the equipment case for signs of damage
7. With all equipment turned on, measure the potential b/w common ground terminal and all conductive surfaces. No surface should be > 5 mV from the ground
8. Periodic inspection of all equipment

21.3 Radiating Equipment

21.3.1 Radiation measurement and detection

The only way to detect and measure radiation is through the use of instruments. There are two categories of instruments used to measure radiation: (i) survey instruments and (ii) dosimeters.

Survey instruments are hand-held, battery-operated devices that measure the rate at which radiation events occur. They can display a 'count rate' expressed in units of counts per minute (cpm); or a 'radiation dose rate' or 'exposure rate' expressed in units of milliRoentgens per hour (mR/h).

In order for radiation to be detected, it must interact with the detector. This produces electrical output signals in the detector which are registered on a scale or ratemeter. The material used as a detector can be a gas, solid or liquid. The most common instrument type is the gas-filled radiation detector. There are many types of gas-filled detectors, but the most common ones used are the ion chamber detector (to measure large amounts of radiation), and the Geiger–Muller (GM) counter (to measure small amounts of radiation). The Geiger–Muller counter is often used in healthcare settings to detect residual radiation. It can only detect the presence of radiation, not distinguish among the types of radiation. With careful calibration and proper use, the Geiger–Muller counter can be successfully used to measure exposures and doses.

A *dosimeter* is a device that measures the total radiation dose over a period of time. This device adds up the entire dose received by a person and provides a cumulative dose result expressed in units of milliSieverts (mSv). A dosimeter should be worn by all personnel who are in frequent close proximity to radiation as determined by the organisation's Radiation Safety Officer. Generally, monitoring is required for those workers who have a reasonable probability of exceeding 25 per cent of the occupational dose equivalent limit of five rem per year.

There are a variety of dosimeters. Examples include film badges, pocket dosimeters, thermoluminescent dosimeters (TLDs), and electronic dosimeters. In Canada, the dosimeter used most commonly to measure external radiation doses is the TLD. Currently, TLDs are the only dosimeters certified by the Canadian Nuclear Safety Commission (CNSC) for external use. They can measure radiation doses between

0.1 m Sv and 10 Sv. There are two types of TLDs – the whole-body and extremity types. Whole-body TLDs are used to measure deep dose and shallow dose over the entire body. Deep dose is the radiation dose received at a depth of 10 mm in the body, while shallow dose is the radiation dose received at a depth of 0.07 mm (skin thickness). Certain handling precautions should be followed with TLDs. These devices should never be exposed to high temperature, water, direct sunlight or fluorescent light. They should be changed in a clean, dry area away from direct light, and skin contact should be avoided. When worn, whole-body TLDs should be attached firmly to clothing between the waist and neck. Extremity

TLDs should be worn facing the radiation source. It is important to never share or wear another person's TLD. If the TLD has been contaminated or lost, the wearer should stop working with radiation until s/he has been issued a replacement TLD. TLDs should be stored in a holder or rack in a low radiation background area when not in use.

21.3.2 Principles of radiation safety

Guidelines for safety are based on three important principles:

1. The longer the exposure, the greater the radiation absorbed.
2. The intensity of the radiation decreases as the distance from the source increases.
3. Specific materials can be used as shielding – the type of shielding material depends on the radiation source.

21.3.3 Effects of radiation on human body

There are two general categories of the effects of radiation on our bodies: (i) heredity effects and (ii) somatic effects.

Heredity effects become apparent in future generations. There has been little research in this area; however, leukemia and mental retardation have been seen in children whose mothers were heavily irradiated while pregnant. For the general public, the documented risk of hereditary ill health in children conceived after either parent is exposed to radiation is estimated to increase by one per cent per 1000 mSv. For workers whose job entails use of radiation, the heredity risk of exposure to radiation is estimated to be 0.6 per cent per 1000 mSv. Why the lower risk among workers than the general public? Generally, workers are only exposed to small measured doses during their working years (between the ages of 18 and 45), while the public can be exposed through therapy, bombings or other sudden events from birth to death.

Somatic effects are those experienced by persons exposed to excessive amounts of radiation. There are two categories of somatic effects:

- *Stochastic effects*: Radiation exposure will increase the probability of developing cancer – the greater the exposure, the more likely that a cancer will develop.
- *Deterministic effects*: Effects of radiation will appear following a threshold exposure. The severity of the effect will depend on the exposure. There are long-term and short-term somatic effects. Cancer is classified as a long-term effect. Examples of cases involving radiation-induced cancer include the early scientists who experimented with radiation, many of whom died from skin and bone cancers and leukemia. Many radium watch dial painters died of bone cancer eight to forty years later. The latency period – delay between exposure to radiation and the onset of cancer – varies with each cancer. Leukemia has a latency period of approximately eight years. There is no scientific evidence to support the existence of a threshold minimum dose below which ionising radiation does not increase the risk of developing cancer. However, the estimated risk for a worker developing a fatal cancer as a result of radiation exposure is an extra four per cent per 1000 mSv. This estimated risk is based on a person working for fifty years and receiving 20 mSv of radiation per year for a total lifetime dose of 1000 mSv. It is important to note that the general public has a 25 per cent risk of developing a fatal cancer in their lifetime; thus, a radiation worker's total risk becomes 29 per cent.

Short-term somatic effects result from both chronic and acute exposures. Chronic exposure is an exposure to low doses of radiation over several months or years. Acute exposure is an exposure to a high dose of radiation within a day. When a person is exposed to very high doses of radiation, there can be significant health effects. Cells most affected include those in the skin, the blood-forming tissues, the gonads and the digestive system lining.

Radiation sickness is caused by exposure of the whole body to high doses of radiation (>2 Gy) causing damage to the body's immune system which impairs the body's ability to fight off infection and disease. Several hours following this high exposure, nausea and vomiting will occur, leading to diarrhoea and general weakness. With even higher doses of radiation (>10 Gy), the intestinal lining is damaged, preventing absorption of water and nutrients. If an individual with whole-body doses near 7 Gy does not receive medical attention, there is approximately 50 per cent chance that s/he will die within 60 days. Doses above 15 Gy can result in damage to the nervous system, which can actually kill a person within a few days. Individuals receiving more than 20 Gy will suffer severe cardiovascular damage and it is likely that 100 per cent of these cases will die.

While this information is certainly frightening, it must be put into perspective. It is important to note that very few people have ever received doses greater than 2 Gy. Furthermore, by following the recommended radiation safety precautions and adhering to the regulatory limits, workers can normally limit their exposure to no more than 0.05 Gy in a year.

Nursing care and safety precautions will depend on the radioactive source, dosage and method of administration of the specific radiotherapy. All healthcare personnel can prevent excessive radiation exposure by ensuring that they are working safely and diligently following all preventive measures. These preventive measures must be clearly documented in the organisation's Radiation Safety Program.

21.3.4 Radiodiagnostic procedures

Preventing radiation exposure

All staff should keep their exposure 'as low as reasonably achievable' (ALARA). Measures to achieve this include:

1. Minimising the time spent in direct contact with the patient – the less the contact time, the lower the exposure.
2. Maximising the distance from the patient – the greater the distance from the patient, the lower the exposure. A distance of three feet is considered reasonable.
3. Use of various shielding materials, depending on the type of radiation.
4. Types of Radiation Hazards

Potential radiation exposure hazards for nursing staff, support staff and visitors who come in contact with radiotherapy patients include:

1. External irradiation by emissions from radionuclides in the patient.
2. Accidental contamination of the skin by radionuclides, leading to ingestion of radioactive material.

Preventing external radiation exposure

Dealing with external radiation is relatively straightforward. Radiation sources can be measured with instruments and exposure stopped by removing yourself from the source. Workers can substantially decrease their external radiation exposure by observing the following precautions:

1. Minimise the time spent near the source.
2. Allow the source to decay before going near it; the radiation intensity will decrease by half for every half-life we stay away from the source. This approach should be used when working with short-lived radionuclides.
3. Maximise the distance between from the source; there is a significant reduction in radiation intensity as distance from the source increases.
4. Use shielding materials e.g., lead for gamma shielding.

Preventing unnecessary contamination of the skin and ingestion of radioactive material

Surface contamination occurs when radioactive materials are deposited on surfaces such as tables, floors, counters, etc, through accidental spills or poor handling techniques. The radioactive material deposited can enter the body in three different ways: inhalation, ingestion and absorption through the skin. For example, a surface contamination can result in contamination of the exposed skin, inhalation resulting from re-suspension of the contaminant, and secondary ingestion exposure occurring when inhaled material cleared from the lungs enter the gastrointestinal tract.

Systemic therapy demands much more stringent precautions than surface contamination. It is important to note that if a radiation source is taken inside the body, all of the radiation emitted is capable of interacting with the body. Also important is the fact that some radionuclides concentrate themselves in certain body organs or tissues. Internal sources continue to irradiate the body constantly until they have been eliminated through excretion and decay. Some radionuclides are eliminated quickly, but others remain in the body for years. Therefore, it is of utmost importance that precautions are diligently followed when handling radionuclides or providing care to patients undergoing radiotherapy.

Preventing radiation exposure

Generally, healthcare personnel may only encounter the possibility of contaminating their skin and subsequently ingesting the source while caring for patients who have been administered volatile radionuclides such as I^{131} or if a sealed source (implant) has ruptured (such rupture, however, is extremely rare). The following precautions should be observed under these circumstances:

1. In the event of spillage of radioactive vomitus or urine, surgical gloves and mask must be donned and an attempt must be made to contain the spill, without directly touching it, by using disposable paper towels or other absorbents. The gloves and contaminated absorbent material must be retained for radioactive monitoring by the Radiation Safety Officer.
2. Urine and feces must be disposed in a toilet. Gloves must be donned to handle the bedpan. The bedpan must be rinsed with water, and the water disposed in the toilet.
3. Good quality disposable gloves made of latex or a strong synthetic material (not vinyl) must be worn whenever there is a potential for contact with radioactive materials. Radioactive materials should never come in contact with the skin. Double-gloving is recommended when there is direct contact with body fluids deemed radioactive.
4. A personal dosimeter must be worn. It must be stored in a place where there is no additional source of radiation.
5. Any clothing soiled with bodily fluids must be removed and retained for radioactive monitoring by the Radiation Safety Officer.
6. Hands must be washed thoroughly following glove removal, after any potential contact, and before eating or engaging in any other hand-to-mouth activity.

To sum up, the following are required:

For radioisotopes or X-ray machines

1. Personnel protective apparel
2. Other protective items (bench paper, shields, segregated storage, etc.)
3. Personnel dosimeters and bioassays, if any
4. Survey meters
5. Routine and special surveys–appropriate frequencies, techniques, instrument, etc.
6. Good handling techniques–remote handling devices, pipettes, shipping vial opening equipment, etc.
7. Good waste handling techniques–segregation, equipment and techniques
8. Discussion of any other pertinent safety aspects of materials or operations, such as carcinogenicity, biohazards, explosives, abnormal fire hazards, etc., which contribute to the hazard of the work.

For radionuclide/radiochemical procedures

Enter μCi and volume amounts in each cell (stock, aliquot, volatiles, wastes, type of experiments and their frequency). Details of the experiments like northern, southern and western blots, nuclear run-on assay, kinase assay, metabolic labelling of protein, receptor ligand binding assay, iodination, DNA sequencing and *in situ* hybridisation techniques are to be mentioned and its potential radiation hazards should be kept in mind.

21.3.5 Radiotherapeutic precautions

Nurses and other healthcare personnel can safely care for patients receiving radiotherapy by following general radiation safety precautions, such as:

1. Reducing the amount of time their bodies or portions thereof are exposed to radiation.
2. Wearing gloves and a lab coat at all times when handling radioactive materials or potentially contaminated materials (IV equipment, patient excretions).
3. Working at the greatest distance possible from the radiation source without compromising patient care.
4. Using shielding whenever possible.
5. Not eating, drinking, smoking, or applying cosmetics in rooms where radioactive materials are present.
6. Posting appropriate signs on patient rooms and restricting personnel and visitors.
7. Practicing standard precautions at all times.
8. Washing of hands following any potential contact.
9. Using positioning and restraining devices to minimise the need for manual patient handling. It is important to keep as much distance as possible between the patient and healthcare personnel – the greater the distance, the lower the exposure rate. It is also important to limit the time spent in direct contact with the patient.

In addition to these general measures, there are specific preventive measures for each type of radiotherapy treatment: sealed source implants, unsealed source therapeutic procedures, unsealed source diagnostic procedures and external radiation beams.

Patients with sealed source implants

Patients who require sealed source implants are generally hospitalised for a few days while the implant is in place. The following precautions should be observed:

1. Patients should be isolated in a separate room, appropriate signs posted, and a lead shield placed in the room. Nursing personnel should work behind the shield as much as possible.
2. Children and pregnant women should be restricted from visiting.
3. The Radiation Safety Officer must post recommended stay times. These should be observed and enforced to prevent undue exposure to radiation.
4. Linen, trash and other patient articles should be left in the patient's room until surveyed by radiation safety personnel.
5. A lead container should be kept in the room for containment in the event that the sealed source becomes displaced from the patient. The source MUST NOT BE TOUCHED with fingers – forceps or some other device must be used in order to keep the source at a minimum of 6–12 inches from the caregiver's hands and body. Appropriate personnel should be notified immediately.
6. Patient's room should not be cleaned by housekeeping or released for use to other patients until the radiation safety personnel have surveyed it.
7. Gynecological patients should be restricted to bed during the therapy treatment. Perineal care is not normally given during the treatment, however, the perineal pad may be changed as and when necessary. If the pad is changed, it is important that the radioactive sources are not disturbed. If perineal care is necessary, safety precautions regarding distance, time and shielding should be observed while this is provided.
8. Good quality gloves made of latex or a strong synthetic material (not vinyl) should be used.
9. Hands should be washed thoroughly following glove removal.
10. A personal dosimeter should be worn as recommended by the Radiation Safety Officer.
11. Dosimeters should be stored in a place where there is no additional source of radiation.

Patients undergoing unsealed source

Therapeutic Procedures Most patients who receive radiopharmaceuticals pose a minimal contamination risk to staff. In general, the use of standard precautions is sufficient to protect them from radioactive blood or body fluids. The only additional precaution to be observed is that of limiting exposure to a developing foetus – thus, pregnant staff should not care for these patients. Other prescribed preventive measures include:

1. Reading and adhering to stay times as recommended by the Radiation Safety Officer.
2. Restricting children and pregnant women from visiting.
3. Treating all objects in the room (including leftover food) as potentially contaminated and wearing gloves when touching anything in the room.
4. Ensuring that all waste is placed into radioactive waste bins and that uneaten food and drink is flushed down the toilet
5. Handling personal protective equipment as radioactive waste when removing or disposing of it.
6. Using good quality gloves made of latex or a strong synthetic material (not vinyl). Double-gloving is recommended when there is direct contact with body fluids deemed radioactive.
7. Washing hands thoroughly following glove removal
8. Wearing a personal dosimeter as recommended by the Radiation Safety Officer.

9. Ensuring that personal dosimeters are stored in a place where there is no additional source of radiation.
10. Patients who have just received I^{131} will contaminate anything they touch immediately following their treatment through their perspiration. In addition, 80 per cent of the iodine may be present in the urine during the initial 48 hours. Therefore, wearing gloves when changing bed linen, touching the client or engaging in any other client care activities is an important precaution for all healthcare personnel.

Patients undergoing unsealed source diagnostic procedures

Standard precautions provide suitable protection from radioactive contamination when drawing blood and collecting urine/faeces samples, as the relative amounts of radioactivity in such samples is quite low. Shielding of these patients is not required as long as all members of the staff are exercising measures to keep their exposure to a minimum through the principle of ALARA. This can be achieved through minimising the time spent in direct contact with the patient - the less time in contact, the lower the exposure. In addition, the distance away from the patient should be maximised - the further the distance from the patient, the lower the exposure. In practice, a distance of three feet is considered reasonable.

Any personnel required to hold children who are given diagnostic doses of gamma-emitting radionuclides must be at least 18 years of age, and the patient should not be held any longer than is necessary during the first 24 hours following the administration of the radionuclide.

Pregnant personnel should limit the time spent in contact with a patient for the first 24 hours after a diagnostic dose of technetium-m99 (^{99m}Tc) and for the first two hours after a gallium-67 (^{67}Ga) injection. Both of these radionuclides have significant gamma and beta energies.

In addition, the following precautions should be observed:

1. Treat all IV and patient fluids as radioactive until proven otherwise.
2. Wear gloves to the maximum extent possible when working with patients and at all times when handling radioactive imaging agents.
3. Store unused radioactive imaging agents in a room separate from the work and waiting areas.

Patients exposed to external radiation beams

No special precautions are required for personnel that are not working directly in the diagnostic imaging/X-ray department. For personnel working in diagnostics, or for those personnel assisting with X-rays/portable X-rays, the following precautions should be observed:

1. Use the shortest exposure time possible that will not compromise patient diagnosis or treatment.
2. Minimise the amount of time that your hands are in an operating radiation beam (e.g., during fluoroscopy).
3. Exit the room, if possible, whenever an X-ray machine or fluoroscopy unit is operating.
4. Wear a lead apron when working around X-ray, CT or fluoroscopy equipment.
5. Wear a personal dosimeter as recommended by the Radiation Safety Officer.

The risk of exposure to radiation is greater during fluoroscopy than during conventional radiographic studies. This is due to the increase in radiation scattering and the increased exposure time. The body areas at greatest risk of exposure for personnel within six feet of the radiation source include the head, neck and hands.

During fluoroscopy, personnel should wear:

i. Lead aprons
ii. Thyroid shields when in close proximity to the patient
iii. Eye protection (e.g., lead-glass lenses) when in close proximity to the patient during frequent or long periods of fluoroscopy
iv. Radioprotective gloves if hands are exposed to the direct beam

Personnel not directly involved in fluoroscopy procedures (e.g., OR attendants, anaesthetists) should also be protected generally by using shielding material.

Organisations that administer and/or use radiotherapy must appoint a competent individual as Radiation Safety Officer (RSO). The primary responsibilities of the RSO are to develop, coordinate and supervise the daily operation of the Radiation Safety Program and ensure protection of all personnel from undue exposure. In addition, the RSO must determine which personnel are in frequent proximity to radiation so that they can be monitored, their exposures recorded, and accurate records maintained.

While the Occupational Health and Safety Act and the Health Care and Residential Facilities Regulation certainly prescribe and delineate roles and responsibilities of workplace parties to develop, implement and maintain a safety program for the protection of workers, only a few countries have specific legislation governing the use of radiation. In Canada the Nuclear Safety and Control Act received royal assent on March 20, 1997 and was promulgated on May 31, 2000. The aim of the Act is to provide for: 3 (a) 'the limitation, to a reasonable level and in a manner that is consistent with Canada's international obligations, of the risks to national security, the health and safety of persons and the environment that are associated with the development of nuclear energy and the production, possession and use of nuclear substances and prescribed equipment and prescribed information. (Nuclear Safety and Control Act)'

Essentially, the purpose of the Act is to provide for the safe use of radioactive materials and radiation devices. The Act applies to all nuclear substances and to all radiation machines that can emit radiation more energetic than 10 million electron volts (10 MeV). It is enforced by the Canadian Nuclear Safety Commission (CNSC).

To sum up:

For Radiotherapy:

1. An isolation ward is usually used.
2. Lead screens may be placed on either side of the bed or at the doorway to block any radiation that is given out.
3. Visitors will be restricted, and allowed to stay in the room or sit at the end of the bed only for a short time, if at all. Visitors will be able to talk to the patient from outside the room through an intercom.
4. Children under 18 and women who are pregnant will not be allowed to visit.
5. Staff and visitors will be asked to stand away from the patient's bed to reduce their exposure to the rays.
6. An instrument called a Geiger counter can be used to monitor the level of radiation in the room. The nurses may wear a small counter.
7. If the patient is breastfeeding, she should stop for the period of treatment and a short time afterwards.

These safety measures and visiting restrictions can make the patient feel very isolated, frightened and depressed at a time when s/he would like people around him/her. Care should be taken to explain to the patient that these restrictions are there only for the safety of his/her near and dear ones.

Interim Life Safety Measures (ILSM) consist of the following actions:

1. Ensuring that exits provide free and unobstructed egress. Personnel shall receive training if alternative exits must be designated. Building or areas under construction must maintain escape facilities for construction workers at all times. Means of egress in construction areas must be inspected daily.
2. Ensuring free and unobstructed access to emergency department/services and for emergency forces.
3. Ensuring that fire alarm, detection, and suppression systems are not impaired. A temporary, but equivalent system should be provided when any fire alarm system is impaired. Temporary systems must be inspected and tested monthly.
4. Ensuring that temporary construction partitions are smoke-tight and built of non-combustible or limited combustible materials that will not contribute to the development or spread of fire.
5. Officers of the Environmental Health and Safety Department or the contractor should provide additional fire-fighting equipment and training on use for personnel.
6. Prohibiting smoking in or adjacent to construction areas in accordance with the University Hospital policy of maintaining a smoke-free environment.
7. Storage, housekeeping and debris removal practices which reduce the flammable and combustible fire load of the building to the lowest level necessary for daily operations must be in place and enforced.
8. A minimum of two fire drills per shift per quarter will be conducted in areas under ILSM.
9. Increased hazard surveillance of buildings, grounds, and equipment with special attention to excavations, construction areas, and construction storage will be conducted by the Environmental Health and Safety Department.
10. The Environmental Health and Safety Department will train personnel when structural or compartmentation features of fire safety are compromised.
11. The Environmental Health and Safety Department will conduct facility-wide safety education programs to ensure awareness of Life Safety Code deficiencies, construction hazards and specific ILSM.

21.4 Further Reading

1. Raja Rao, C. and Guha, S. K. *Principles of Medical Electronics and Biomedical Instrumentation.* Universities Press (India) Ltd. . 2001.
2. Khandpur, R. S. *A Handbook of Biomedical Instrumentation.* Tata McGraw Hill. 2002.
3. www.stanford.edu/dept/EHS/prod/researchlab/radlaser/ manual/appendices/forms/declaration.htm
4. www.stanford.edu/dept/EHS/prod/researchlab/ radlaser/manual/part3/humanuse.htm
5. oregonstate.edu/dept/radsafety/procedures/ rsc102_amendment_application.doc
6. radiationsafety.asu.edu/X-ray/chap_5.cfm
7. www.ehs.stonybrook.edu/health/hc-iispn.asp
8. collection.nlc-bnc.ca/100/200/300/health_care_health_safety_assn_on/ safety_consideration_health/rrade334.pdf
9. Radiation safety considerations in the management of patients receiving therapeutic doses of 131-I. *Operational Health Physics.* Proc. 9th Midyear Topical Symposium of the Health Physics Society pp. 279–286.1976.
10. Internal contamination in nurses attending patients that received therapeutic amounts of radioiodine-131. W. Termoshuizen et al. IRPA Conf. Proc. p.747. Sidney.1988.

11. Precautions in the management of patients who have received therapeutic amounts of radionuclides, *NCRP Rep.* #37. 1970. Radiation Safety Institute of Canada 2002. Radiation Safety Training Course 2nd Edition. Toronto
12. American Brachytherapy Society www.americanbrachytherapy.org/open.htm
13. CancerBACUP (UK) – Understanding Radiotherapy www.cancerbacup.org.uk/info/radio/radio-7.htm
14. Cancer Index www.cancerindex.org
15. Duke University Hospital – Nursing Radiation Safety Page www.Safety.duke.edu/RadSafety/nurses/default.asp
16. Frequently Asked Questions About Radioactive Patients on Duke Hospital Divisions www.Safety.duke.edu/RadSafety/info_materials/FAQsheet2_Nursing.pdf
17. Nuclear Medicine Research Council www.cbvcp.com/nmrc
18. Penn State Hershey Medical Center Radiation Safety Manual – Nursing Procedures www.xray.hmc.psu.edu/services/hphys/rsm/nurseproc.html
19. Radiation Safety Institute of Canada www.radiationsafety.ca Radiation Therapy Oncology Group www.rtog.org
20. Society for Nuclear Medicine: About Nuclear Medicine – Patient Information www.snm.org/nuclear/new_patient_info_1.html

21.5 Questions

1. Descriptive Questions

1. Why do you think safety measures are important for the various types of electromedical equipment?
2. What are the differences in the safety requirements for radiotherapy vis-à-vis radiodiagnostic equipment?

2. Write Short Notes on

1. Preventive measures for electrical shock hazards from biomedical equipment.
2. Precautionary measures for radiotherpaeutic procedures.

3. Multiple Choice Questions

Choose the single best response.

1. The most lethal range for electrical shock is
   (A) 100–200 mA
   (B) 50–100 mA
   (C) 200–300 mA
   (D) none of the above

2. Macroshock refers to electrical current shock applied to
   (A) a small area externally
   (B) a large area internally
   (C) a large area externally
   (D) a small area internally

3. Which statement is incorrect: Ionizing radiation at very high doses causes the cells
   (A) either to die or to be changed permanently.
   (B) the body does not replace the cells that die, causing ill effects.
   (C) the cells that have permanently changed produce abnormal cells and
   (D) under the right conditions, these abnormal cells can become cancerous.

4. For radiotherapeutic procedures, the true statement is:
   (A) Children under 18 and women who are pregnant are not allowed to visit the patient
   (B) Staff and visitors are asked to stand away from the bed to reduce their exposure to the rays
   (C) If the patient is breastfeeding, she should stop for the period of the treatment and a short time afterwards
   (D) all of these

5. The principles for radiation safety are:
   (A) The longer the exposure, the greater the radiation absorbed
   (B) The intensity of the radiation decreases as the distance from the source increases
   (C) Specific materials can be used as shielding – the type of shielding material depends on the radiation source
   (D) all the above

Answers

1. A; 2. C; 3. B; 4. D; 5. D

22

Planning biomedical instrumentation for diagnosis, therapy and rehabilitation

22.1 Introduction

Whenever technology, especially a new one, is to be introduced in the running setup of a healthcare delivery system (either a tertiary care superspecialty referral hospital or a remote rural primary health sub-centre), careful planning is of utmost necessity. As you can easily guess, the requirements and financial capabilities of the extreme ends of these spectra are very much different.

However, appropriate planning will be similar in both the cases.

22.2 Strategic Planning Process

One such planning methodology is termed as the *strategic planning process*, which at a reasonable cost

i. integrates technologies of diverse nature, but are related to the area(s) of interest pertaining to that particular centre;
ii. and ensures high quality care to all the patients.

Citing a concrete example would be more comprehensible. Suppose, a cardiac care centre (e.g., having intensive care) decides to install some new facilities. If there is no future plan of extension into a multispecialty centre, buying an MRI scanner may be a total waste. Similarly, a specialty hospital of neurosciences (medicine and surgery) may not require a treadmill as much as a cardiac centre would do. Therefore, big hospitals often prefer to diversify so that the expensive resources are utilised optimally. Other than in cardiology, a treadmill may then be used for physical and sports medicine too.

Now, simply buying a machine (even if we have enough money) is not the end of the story, rather it is the beginning. Most costly equipment would require some form of training to the personnel responsible for its day-to-day operation and maintenance. Often, a trained person may be used for some other job (which may require much less skill). Therefore, it then becomes a man-management problem, rather than a technical or financial one. And, unless someone is held responsible and accountable for the functioning and proper utilisation of the equipment, often the machine breaks down due to inadequate or faulty use.

One word of caution should be kept in mind. It is not always that a costly equipment or procedure will yield better quality patient care. Organisations like the ARSI (Association of Rural Surgeons of India, see Chapter 23) are building their sophisticated surgical instruments locally.

This naturally leads to the question of affordability for the masses. Some private hospitals have the best equipment in the world, yet the charges are beyond the reach of the common man. Some argue whether one should think of the cost or the quality of life after the procedure. We shall deal with these issues in the next chapter.

22.3 SWOT Analysis

A well-known method for strategic planning is performing a *SWOT analysis*, where,

S Strengths: The strengths of the organisation in its current form have to be listed first. The strength may lie in highly motivated and well-trained personnel, or an excellent management or goodwill or huge financial resources.

W Weaknesses: The organisation's weaknesses in its current form have also to be enumerated. It may lie in a bad work culture, lack of adequately trained personnel, directionless management or pecuniary constraints.

O Opportunities: The organisation's opportunities also need to be listed. They should be as follows: (A) External opportunities, e.g., in the form of a large number of satisfied patients motivating other patients to come; (B) Internal opportunities e.g., a great number of happy employees may encourage their highly skilled friends to join the organisation.

T Threat: The threats to the organisation also have to be carefully assessed: (A) External threats: may be in the form of external agitation or fierce competition with other efficient healthcare providers; (B) Internal threats: could be internal agitations and sabotage for small gains.

Once the SWOT analysis is done impartially, the results can decide the future course of action. If the manpower (human resources) seem to be skilled, satisfied and sufficient, almost any decision is likely to yield very good results. On the other hand, if these (in the form of inner strength) is lacking, introducing even a small load of extra work may paralyse the whole setup. Also it is essential to take all people (not just employees, but end users like the patient and their family members) into confidence when major decisions are to be taken by the organisation.

However, this team spirit is lacking in most failing healthcare delivery systems.

22.4 Life-Long Learning

In today's global village, everything seems to change almost daily, especially if it is in anyway linked with technology. To cope with this change, one has to keep oneself very much up-to-date. The term

i. Determining how the deployment should be evaluated, in context of the specific healthcare delivery centre.
ii. Assisting in the review of emerging technologies and their impact, especially on the life of neighbouring people.
iii. Articulating justifications and provisions for adoption of technologies, training and retraining all concerned personnel.
iv. Considering visits to research laboratories and exhibits at major meetings and trying to practically use the knowledge so gained.
v. Gaining familiarity with the institution and its equipment users' ability to assimilate new technology, and motivating them.
vi. Meeting, and better still, if possible, exceeding, standard of medical care.
vii. Impacting care quality or level.
viii. Impacting life quality.
ix. Improving intervention's accuracy, specificity, reliability, and/or safety.
x. Increasing rate of recovery.
xi. Catering to the needs and desires of the community at large, and more so to whom the services are provided directly.

22.5 Planning for Various Facilities

We will not be discussing detailed planning process for the various types of facilities in this book. Although a brief overview is presented below.

Inpatient hospital

A facility, other than psychiatric, which primarily provides diagnostic, therapeutic (both surgical and non-surgical) and rehabilitation services by, or under, the supervision of physicians to patients admitted for a variety of medical conditions. Therefore, this would require the maximum attention and planning. In the case of a psychiatric hospital (earlier known as mental asylum or lunatic asylum), a more thorough and holistic approach is required. Most tertiary care hospitals like medical colleges or private superspecialty hospitals fall in this category.

Outpatient hospital

A portion of a hospital that provides diagnostic, therapeutic (both surgical and non-surgical), and rehabilitation services to sick or injured persons who do not require hospitalisation or institutionalisation. Almost any inpatient hospital has an outpatient department (OPD). In such hospitals, the basic infrastructure is already present. However, there are some clinics which provide only outpatient facilities. These are mostly located either in remote areas or in some residential campus. The requirements are fewer than in an inpatient hospital but the number of patients is far more. Therefore, all the facilities (including human resources) must be adequate in quantity.

Emergency room hospital

A portion of a hospital where emergency diagnosis and treatment of illness or injury is provided. Usually there are no isolated emergency or trauma care centres. They are generally part of a big tertiary care institute where highly-skilled personnel and sophisticated facilities are easily available. The most important planning requirement in these centres is in the checking of functionality (expiry dates of drugs, emptiness of oxygen cylinders or blocking of the sucker tubes). Someone has to be responsible and accountable for the proper functioning at the time of need. An excellent example in our country is the ARME (Accident Relief Medical Equipment) coach of the Indian Railways, available at all divisional headquarters and important stations.

Ambulatory surgical centre

A freestanding facility, other than a physician's office, where surgical and diagnostic services are provided on an ambulatory basis. Many NGOs (non-governmental organisations) have such services throughout our country. These centres provide a mixture of OPD and emergency care. Nowadays, they are often linked to telemetry devices and sent for remote teleconsultation with specialists sitting in bigger cities. In some places helicopters and airplanes are also used for this purpose. Planning for this facility must take into consideration all these factors.

Comprehensive inpatient rehabilitation facility

A facility that provides comprehensive rehabilitation services to inpatients with physical disabilities under the supervision of a physician. Services include physical therapy, occupational therapy, speech pathology, social or psychological services, and orthotics and prosthetics services. This has now become an integral part of rehabilitation medicine. Again, most tertiary care hospitals have such facilities. This is indeed a place where multidisciplinary interaction is not just a phrase but a way of life. Appropriate teamwork is absolutely necessary for the proper functioning of such a unit. Therefore, the planning has to be elaborate and in consultation with all concerned.

Comprehensive outpatient rehabilitation facility

A facility that provides comprehensive rehabilitation services under the supervision of a physician to outpatients with physical disabilities. Services include physical therapy, occupational therapy, and speech pathology services. There are various specialised centres for the physically (mentally, visually or hearing) challenged persons being retrained to perform various daily functions as well as vocational and educational activities. In these centres the planning should not only stress on newer equipment but also on adequately trained and compassionate manpower (often ladies fit in such roles much better).

22.6 Further Reading

1. Raja Rao, C and Guha, S. K. *Principles of Medical Electronics and Biomedical Instrumentation.* Universities Press (India) Ltd. 2001.
2. Khandpur, R. S. *A Handbook of Biomedical Instrumentation.* Tata McGraw Hill. 1987.
3. Redmond, B. and Richardson, V. Just getting on with it: Exploring the service needs of mothers who care for young children with severe/profound and life-threatening intellectual disability. *Journal of Applied Research in Intellectual Disabilities* 16: 205–218. 2003.
4. www.spastic.org.sg/

22.7 Questions

1. Descriptive Questions

1. Why do you think that proper planning is mandatory for setting up an instrumentation laboratory with respect to healthcare delivery ?

2. Write Short Notes on

1. SWOT Analysis
2. Life-long learning for bioinstrumentation.

3. Multiple Choice Questions

Choose the single best response.

1. Rehabilitation falls under the category of ___ prevention
   (A) primary
   (B) secondary
   (C) tertiary
   (D) none of these
2. Therapeutic planning includes
   (A) surgical
   (B) medical
   (C) radiation
   (D) all
3. SWOT analysis gives an idea of
   (A) only negative aspects
   (B) only positive aspects
   (C) an overall view
   (D) none of the above
4. Strategic planning process includes
   (A) affordability
   (B) accessibility
   (C) accountability
   (D) none of the above
5. Life-long learning is
   (A) a very new concept
   (B) a traditional concept
   (C) related to information technology
   (D) only a fancy word

Answers

1. C; 2. D; 3. C; 4. A; 5. B

23

Social and Ethical Issues related to Biomedical Engineering

23.1 Introduction

Medical ethics or *bioethics* is nothing but the study of moral issues in the fields of medical treatment and research. The term is also sometimes used more generally to describe ethical issues in the life sciences and the distribution of scarce medical resources. There have been several early codes of ethics, including the Code of Hammurabi (the oldest recorded one). The next was the ancient Greek Hippocratic Oath, which required physicians above all to 'do no harm'. The professional codes of ethics written by English physician Thomas Percival in the 18th century provided a foundation for the first code of ethics established in 1846 by the founders of the American Medical Association. It was none other than Adolf Hitler who indirectly showed the importance of moral ethics in medical practice and research. The infamous 'human' medical experiments in the Nazi concentration camps forced into existence the Nuremberg Code for ethics in research on human subjects. The code was established during the war crime trials at the close of World War II (1939-1945). The advent of new medical and reproductive technologies after the 1950s further complicated the moral and societal issues of medical research and practice. The historical background of social and ethical issues in the modern healthcare delivery sector is enlisted below.

1. Nuremberg code (1947): voluntary consent
2. Universal Declaration of Human Rights (UN General Assembly, 1948): concern for human subjects being subjected to involuntary maltreatment

3. Helsinki Declaration (Council for International Organizations of Medical Sciences, CIOMS, 1964): modified from time to time.
4. International Covenant on Civil & Political Rights (1966): No one shall be subjected to torture or to cruel, inhuman or degrading treatment or punishment. In particular, no one shall be subjected without his consent to medical or scientific treatment.
5. Policy Statement on Ethical Considerations involved in Research on Human Subjects (ICMR, 1980): one of the earliest documents for biomedical research ethics.
6. Proposed International Guidelines for Biomedical Research involving Human Subjects (WHO, CIOMS, 1982)
7. International Guidelines for Ethical Review in Epidemiological Studies (CIOMS, 1991)
8. International Ethical Guidelines for Biomedical Research involving Human Subjects (CIOMS, 1993)

23.2 Principles of Clinical Research

The Central Ethics Committee on Human Research (CECHR) of ICMR has proposed the Ethical Guidelines for Biomedical Research on Human Subjects. In its full form, it is freely available online at www.icmr.nic.in website.

The guidelines have raised some specific issues of concern in the respective 'hot' areas of biomedical research:

i. Genetic Engineering
ii. Assisted Reproductive Technology (ART)
iii. Stem-cell Research
iv. Human cloning

New genetic technologies promise to make medical ethics an even more central part of social decision-making. The Human Genome Project, a 15-year, federally funded $3 billion effort to code the entire human genetic map, has already resulted in the discovery of a number of genes that may lead to particular diseases or traits. This project is also giving individuals more information about their own genetic make-up. Medical ethicists are debating whether or not this genetic information is the exclusive property of patients, or is instead properly the concern of insurers, employers, and society.

Gene therapies are also being developed that use genetically engineered viruses to manipulate the cells of patients. Some wonder if the manipulation of human cells through genetic engineering is somehow contrary to the laws of nature or religion; others have proposed that it will lead to the manipulation of human sperm or eggs for eugenic purposes (improving the hereditary qualities of a race, as widely practiced in the Nazi regime).

Determining the sex of the unborn baby and eliminating 'unwanted' pregnancies with that knowledge can also be easily practiced using ART. This leads to a 'gender bias' in society and is a matter of grave concern to society.

Albeit stem cells (pluripotent or having the capacity to become any type of specialised cell of the body like bone, muscle, nerve, liver or blood vessels) are essential for giving life to many patients and replacing the need for an organ transplant, the method of harvesting and wasting the unused stem cells raise serious ethical questions.

Cloning, the production of organisms genetically identical to a parent, has also become a controversial topic in medical ethics. In 1997 Scottish scientist Ian Wilmut and his colleagues announced the birth of a sheep named Dolly that was produced from a cell of an adult female sheep. Now the grey area is the ethics of creating human clones, the circumstances under which human cloning might be used, and the possibility of using the technique to manipulate the traits of children. This issue, as most other ethical issues, remains unresolved and will continue to challenge medical ethicists well into the 21st century. The birth of a human clone 'Eve' from her mother's cells, claimed by members of the Raelean sect in December, 2002 is still an unsettled fact.

In light of the above discussion, any proposed Biomedical Research Plan's General Statement should have a clear-cut definition of its purpose, conduct and evaluation criteria.

General principles should be strictly followed for:

i. Essentiality
ii. Volition, informed consent, and community agreement
iii. Non-exploitation
iv. Privacy and confidentiality
v. Precaution and risk minimization
vi. Professional competence
vii. Accountability and transparency
viii. Maximisation of public interest and distributive justice
ix. Institutional arrangements
x. Public domain
xi. Totality of responsibility
xii. Compliance

23.3 Randomised Controlled Trials

The accepted protocol for any experimental design is the randomised controlled trial (RCT). The steps for conducting an RCT are discussed below.

23.3.1 Protocol design

The initial process is to design the experiment in the following manner.

1. Drawing up a *protocol* (e.g., *concurrent parallel*, where two groups undergo two different interventions throughout the period of trial or *cross-over*, where the two groups interchange the interventions after a specified period of time, to avoid biases)
2. Selection of a suitable *population* (reference or target) for assessing the efficacy of the proposed interventions.
3. Selection of a suitable *sample* (experimental or study) since experimenting on the whole population will be out of question. The inclusion criteria for the experimental groups have to conform to standard definitions, e.g., from ICD-10 (WHO). The definitions of population and sample are given in Chapter 20.
4. Making necessary *exclusions* (ineligibility/non-consent) from the sample, for finalising the subjects for the trial.
5. *Randomisation* (experimental/control) is absolutely necessary to avoid bias in selection of sample.

6. *Manipulation/Intervention* and follow-up is essential for measuring the effects of the measures applied to the sample population.
7. *Assessment/Evaluation* of outcome has to be carried out meticulously, applying the appropriate statistical measures.

All these steps must be well-defined beforehand, with necessary back-up measures, wherever required.

23.3.2 Types

The RCT may be either: (A) open, (B) single blind, (C) double blind or (D) triple blind.

In the case of open trial, both the investigator and the subject know about the interventions. Therefore, it is likely to be biased and is better avoided unless inevitable. In single blind trials the subject does not usually know of the intervention but the investigator is aware of it.

In double blind trials neither the subject, nor the investigator knows of the intervention. So, it is likely to lead to a rather unbiased conclusion. This is the most common form of clinical trial practiced.

However, the best method is the triple blind trial. Here, apart from the subject and the investigator, the person doing the statistical analysis is also unaware of the identity of the intervention applied to a particular subject. Thus, there is practically no chance of any observer bias.

Some examples of RCT:

i. Clinical (drugs/devices/operative)
ii. Preventive (vaccines)
iii. Risk factor (cholesterol)
iv. Cessation experiment (smoking)
v. Etiological (retrolental fibroplasia in hyperbaric oxygenation to preterm babies)
vi. Evaluation (DOTS or directly observed treatment – shortcourse for tuberculosis)

23.3.3 Risk–benefit ratio

Risk–benefit ratio (RBR) may be defined as the relation between the risks and benefits of a given treatment or procedure. Institutional Ethical Committees (IECs) (located where the study is to take place) determine that the risks in a study are reasonable with respect to the potential benefits. It is also up to the patient to decide if it is reasonable for him or her to take part in a study.

Favourable risk–benefit ratio

Every medical research project involving human subjects should be preceded by careful assessment of predictable risks and burdens in comparison with foreseeable benefits to the subject or to others. (Declaration of Helsinki. Principle 16)

The benefits, risks, burdens, and effectiveness of a new method should be tested against those of the best current prophylactic, diagnostic, and therapeutic methods. (Declaration of Helsinki. Principle 29)

23.4 Technology and Community

Rehabilitation engineering

Rehabilitation engineering is the process of helping people with disabilities achieve and maintain maximum physical function, live in their own homes, attain gainful employment, and participate in and contribute to society.

Independent living recognises that each person has the right to independence through maximum control over his or her life, based on an ability and opportunity to make choices in performing everyday activities. These activities include: managing one's personal life; participating in community life; fulfilling social roles, such as marriage, parenthood, employment, and citizenship; sustaining self-determination; and minimising physical or psychological dependence on others. Community integration incorporates ideas of both place and participation, so that a person is physically located in a community setting, and participates in community activities. Issues of consumer direction and control also are integral to concepts of community integration. The goals of many research programs are to encourage independent living and community integration, to achieve more successful outcomes for people with disabilities, and to foster the development of innovative methods to achieve these outcomes and to measure achievement.

Related disability research emphasises knowledge areas that are cross-cutting and essential to the support and refinement of disability research generally. The common theme linking disability statistics, outcome measures, disability studies, rehabilitation science, and international activities, is that they all provide essential frameworks and building blocks that enable the disability research enterprise to thrive and to address important issues in meaningful ways.

Dissemination and utilisation are the tools through which people with disabilities can become fully integrated and participating members of society. The dissemination and utilisation efforts ensure the widespread distribution, in usable formats, of practical scientific and technological information generated by research, demonstration, and related activities. Our challenge is to reach diverse and changing populations; to present research results in many different and accessible formats; and to use the technology appropriately

The Americans with Disabilities Act (ADA) opens more opportunities for persons with disabilities. It also places certain responsibilities on employers, transit and communication systems, state and local governments, and public accommodations. To assist covered parties to understand and comply with the ADA, NIDRR has funded a network of grantees to provide information, training, and technical assistance to businesses and agencies with duties and responsibilities under the ADA

Augmentative communication

Augmentative communication refers to any method of communication besides speech. People who cannot speak interact more easily with others using a variety of (non-verbal) communication aids. These may include low-tech options such as a manual communication board to a high-tech electronic/computer-based system with synthesised speech output. These devices can provide a person with a 'voice' to express themselves in all areas of their lives. There are alternative methods to help someone with minor speech difficulties as well as those with severe difficulties.

Electronic aids to daily living refer to those devices which augment a disadvantaged person's ability to control common household electronic devices. For persons with limited mobility or hand control, operation of TV/VCR remotes, turning lights on/off or using a telephone is often difficult. Environmental control units enable people to operate these devices with as little as a single switch. As a result, clients are able to live more comfortably and independently.

Computer access refers to a person's ability to utilise a standard computer for business, educational or personal purposes. Individuals who cannot use a standard keyboard or mouse due to limited hand use can access computers in alternate ways. A variety of alternative access methods are available such as miniature or enlarged keyboards, head controlled systems, voice-activated systems, alternative mouse input devices, or specialty software.

Rural surgery

While there are five-star hospitals in large cities serving a comparatively small section of society, there are rural hospitals in the country-side serving the majority. It is this majority population that yields 80% of the country's GDP. It is a pity that 80% of this GDP is consumed by 20% of the upper-crust population. This population is guided by the dictates of the western healthcare industry which is bent on maximising profit. Norms set by them are more bent towards this direction rather than to the benefit of the common man of our country.

Therefore, ideally, while making guidelines, the limited resources of our country has to be kept in mind. Our per capita GNP is just 400 US Dollars while that of western countries vary from 25000 to 40,000 US Dollars!

According to the Voluntary Health Association which had made a countrywide survey in 2001, 400 million people have no access to basic healthcare facilities! Also our country has an average of 0.8 beds per thousand while the WHO recommendation is 4 beds per thousand.

Considering these factors, guidelines should be made with great caution. This guideline should not become a deterrent to a young surgeon aspiring to serve the rural population of our country.

The famous scientist Karl Popper had said (1984) that 'the most important quality of a scientist is intellectual humility'. Those who make the guidelines for standards should have the humility to learn from multi-centric studies of both rural and teaching hospitals as to the minimum and maximum requirements for surgical services.

To do this, our approach should be based on our societal needs and sustainable by the consumers, be it private, voluntary or government. The healthcare economics has to be kept in mind all the time. The surgical care may be divided into the following:

i. *Primary surgical care* those involving minor office procedures like stitching, dressing and draining of minor abscesses.
ii. *Secondary surgical care* those involving routine surgery like appendectomy, cholecystectomy, caesarian section, perforations and obstructions and simple fracture reductions. In other words, those being done in the rural hospitals of our country.
iii. *Tertiary surgical care* those that are being done in teaching institutions including complicated surgical procedures, transplants and cardiac surgery.

ASI (Association of Surgeons of India) and ARSI (Association of Rural Surgeons of India) may evolve a proper referral system across the entire country between the rural surgeons, the impoverished district hospitals, and the teaching institutions. As of now, referrals are made by personal contacts. This does not work all the time and the patient suffers in the middle. If we consider the fact that all surgeries are done for the benefit of the patient and not to glamourise ourselves, there is no reason why it cannot be done on a national scale at optimal cost..

Practical technology options for rural health care are increasing due to

i. Better (more powerful, smaller, cheaper) tools being available (*cf.* locally made ones, e.g., by ARSI)
ii. Internet access (more useful information and interaction)
iii. Better connectivity (telemedicine)

PHA (People's Health Assembly 2000) felt that:

i. There is need to build an integrated movement for people-centred healthcare as opposed to the profit-driven global health industry.

ii. The community must be a participatory actor in the health system that is designed for and directed towards it.

iii. The southern Indian state of Kerala though considered poor in terms of traditional economic indicators such as per capita income and industrial investments, is yet the richest state in India in the context of social development indicators such as healthcare and education. It achieved this not by conventional methods but through mobilisation of the masses at the grassroot level. The gains of the 'Kerala model', which has been praised throughout the world, however is under threat from cuts in subsidies for services such as healthcare under pressure from international funding agencies.

James Orbinski of the Nobel Peace Prize winning *Medecins Sans Frontieres* (Doctors without Borders) said that to begin a genuine global movement for primary healthcare a clear vision, sense of action and voice is needed. 'Today we substitute charity for duty.' he said, 'This is simply not good enough.' NGOs he said have accepted this charity and not fought hard enough for change.

A study conducted by the Centre for Research in Rural and Industrial Development (CRRID), Chandigarh, conceded that women village health guides (VHGs) did play a role, inter alia, in educating ruralites on immunisation, family planning, health-related matters, drinking water, etc. Why then abolish the scheme as some negative minds suggest? Better spread it countrywide. Construction, not destruction, holds a promise for the village India which Gandhiji called 'Real India'.

We shout 'primary health', but provide 'secondary care'. The Association of Rural Surgeons of India reveal that 45 per cent of the surgeons work without an anesthetist, 68 per cent without a radiologist, 68 per cent without a qualified pathologist, 63 per cent without blood bank facilities and 32 per cent do not have any of these things.

According to a medical expert, at least 104.06 million people are suffering from major diseases. Nearly 50 million deaths are caused by water-borne diseases. Official figures hide more than they reveal. Even so, the statistics provided by the Ministry of Rural Areas and Employment are telling. On April 1, 1997, Of the 13,18,899 habitations surveyed, 1,17,429 remained without water, while 3,83,106 received only a part of the total need of water supply. In 16 major states, for which survey reports have been received, 55,739 habitations have brackish water, 58,325 have excessive iron, 37,760 have an excess of nitrate, 27,895 have too much fluoride. This is borne by Dr A.K. Tharian's finding that 25 per cent hospital beds are occupied by patients suffering from water-borne diseases.

The other areas of concern due to technology inroads into the society have been:

i. Multi-disciplinary interaction, e.g., ethnobotany (preservation of the proprietary rights of the traditional herbal healers by drug companies exploiting their knowledge), ergonomics (proper interaction between man and machine optimising the utilisation of both human and material resources), behavioural medicine (psychologist, psychiatrist and medical expert along with occupational therapist planning proper rehabilitative measures for the patient).

ii. Biodiversity preservation versus GMOs (genetically modified organisms).

iii. Information confidentiality in Electronic Health Records.

23.5 Environmental aspects of technology related to healthcare delivery

Considering the role of environmental effects (both biological and psychosocial) of present day technological advances includes:

i. *Bioengineering*: bioprocesses for the treatment of industrial wastewater, bioremediation of groundwater contaminated with chlorinated solvents and other recalcitrant contaminants.
ii. *Microbiology*: biodegradation of pollutants, molecular diagnostics, biotreatability and biomonitoring of contaminated sites or transformation processes of organic residues.
iii. *Ecotoxicology*: toxicological effects of pollutants, development of bioassays, environmental risk assessment.
iv. *Genetics*: catabolic pathways of pollutant biodegradation (aromatics, PCB or polychlorinated biphenyls, etc.), *Bacillus thuringiensis* (biopesticides), mode of action, DNA synthesis and sequencing, development biochips for the detection of pathogens in the environment.
v. *Analytical Chemistry*: identification of contaminants and degradation products in soil, water and biomass, method development, environmental biochemistry.
vi. *Biosensors/Nanobiotechnology*: detection of microorganisms, development of biosensors for various contaminants such as PAHs (poly-aromatic hydrocarbons), chlorophenols, resin acids, and explosives.
vii. *Bioconversion and Sustainable Development*: using functional capabilities of microbes and natural microbial communities for the development of bioconversion and biocatalytic processes in value-additions and bioproducts

23.6 Healthcare economics and health rationing

There is no need to elaborate on the fact that our healthcare delivery resources are rather inadequate. Targeting 'Health For All' was a failure by 2000. The reasons may be that there have been too many dissipated health programs (each having their own goal under the definition) imposed by the centre (and gladly accepted by the states because of the money that comes with it) without understanding the ground realities, especially the healthcare demands of the population. Unfortunately, the state does not insist on meeting all these program objectives; the idea is not to achieve the objectives but to create the impression that the state is benevolent and is doing its best to serve the people through a large array of programs. This image building is not unique to the health sector but also holds good for education, social welfare and rural development programs. Each FYP (five year plan) keeps adding to the list of programs with the result that we have now lost count of the number of programs. The consequence of all this has been a bulging bureaucracy.

On the one hand, the state talks of integration and on the other, it keeps delivering new babies (programs) which develop their own vested interest. (No wonder we still have program staff in our health program for small pox and plague, both of which were eradicated long ago!). All this goes under the name of primary health care.

The other villain under India's primary healthcare program is family planning (FP). The Indian State is sold over to the ideology that all our economic ills, poverty and what have you, are due to the population's rapid, unchecked growth. Hence, family planning becomes the core program under public health services. The entire state health machinery is geared towards achieving the FP targets, no matter what happens to the other health programs.

Another problem with primary healthcare in India is that it is associated only with rural areas. Urban areas have the privilege of enjoying medical care but rural areas will get only primary healthcare. This dichotomous approach hence creates the impression that primary healthcare excludes medical care or curative services. It is all this that has earned primary healthcare a bad name in the country.

In functional terms primary healthcare as practiced in India is implemented through Primary Health Centres (PHC), one each for 30,000 persons (only in rural areas) with six beds for the never-turning-up maternity and emergency cases, one to two doctors and a few inadequately trained and overburdened paramedics who are to provide the services under the various national health programs. It goes without saying that the entire PHC (and its subcentres) staff is engaged in meeting FP targets. The doctors do run OPD clinics at the PHC but the supplies and services are so inadequate that the clientele is automatically restricted in numbers. Hence it is not surprising that the HSSO in its 42nd Round Survey on healthcare utilisation found that of all routine ailments treated in rural areas only 5% were treated in PHCs. The remaining either went to city public hospitals and dispensaries (20%) or to private practitioners (59%) and private hospitals (16%).

The resource constraint theory was an eyewash because the resources needed for the primary healthcare component of the Bhore (the Health Survey and Development Committee, chaired by Sir Joseph Bhore, an Indian Civil Servant, in 1946) proposals was only one per cent of the GDP then; another one and a half per cent would be needed for the secondary and tertiary care components (including medical/nursing/paramedic education). However, our planners and policy-makers thought otherwise and developed the secondary and tertiary care sector, ignoring primary healthcare completely. Thus today we have well-developed city hospitals and a massive medical education infrastructure, which mainly produces doctors for the market. In contrast, the rural areas have underdeveloped PHCs which meet the priority needs of the state and not the demands of the population.

The average medical expenditure (such as fees, medicines, clinical and diagnostic tests, surgery, and hospital bed charges) per episode for inpatient care in 1993 was Rs. 850 for rural and Rs. 1,065 for urban patients; the respective figures for outpatient care were Rs. 70 and Rs. 97. There were large inter-state variations in the cost of treatment. As expected, the cost of treatment was higher in urban than in rural areas, in the private than in the public sector and for inpatient than for outpatient care. For both inpatient and outpatient care, the private sector agencies, on an average, charged three to four times that the public sector agencies. Overall, the medical expenditure constituted 84% of the total cost of treatment; the remaining expenditure included expenses on transport, special diet, rituals, gifts and tips, etc. In most states, the proportion of indirect cost (mainly transportation) was higher for rural than for urban patients, thereby reflecting the poor distribution of healthcare facilities in rural areas especially in hilly states.

On an average, an Indian household spends Rs. 250 per capita per annum on healthcare; the figure for urban households is about 40% higher than their rural counterparts. Nearly 80 per cent of the total health expenditure incurred by households is on private healthcare facilities. This is largely because private healthcare expenditures, both for inpatient and outpatient care, are considerably higher in terms of out-of-pocket payments. Even so, public healthcare services are not free and people do incur considerable out-of-pocket expenditure on public healthcare facilities. The NSS data reveal that more than two-fifths of inpatients and one-third of outpatients using public facilities have to pay for public services. Nevertheless, public facilities are relatively cheaper than private facilities.

There are large inter-state variations in both government and household spending on health care whether expressed in per capita terms or as a percentage of the state domestic product (SDP). The hilly states of Jammu and Kashmir and Himachal Pradesh, as well as Kerala, spend more than nine per cent of their SDP on health whereas it is around three per cent for the relatively developed states of Punjab, Haryana and Maharashtra. It appears from selected healthcare indicators across major states that with the exception of hilly states, government spending on health among the poorer states is low and, therefore, the out-of-pocket expenditure by households in these states is relatively higher. Also, the higher level of morbidity, results in further raising the share of household expenditure in total health spending.

The quality of life (QoL) relates to communities, families, and individuals from a variety of population groups. QoL is an area of study that has attracted a great deal of interest over the past ten years, particularly in the areas of health and social services, but increasingly in medicine, education, and others. The study of QoL is an examination of factors that contribute to the goodness and meaning of life, as well as people's happiness. It also explores the inter-relationships among these factors. The ideological thrust of quality of life study is to promote means for people, within their environments, to live in ways that are best for them. The ultimate goal of quality of life study and its subsequent application to people's lives is to enable people to live quality lives - lives that are both meaningful and enjoyed.

Another indicator of the QoL is the Physical Quality of Life Index (PQLI). PQLI of India was 43 in 1979, while those of Sri Lanka and Korea were 82 each during the same period. This includes GNP, life expectancy at 1 year, infant mortality rate (per 1000 live births) and literacy rate (% of people above 15 years of age).

The debate on the quality of life versus the sanctity of life continues throughout the world.

Health Rationing

Health rationing may be defined as: To ration health care is to withhold at least some medical services from at least some individuals who would probably benefit from them, because we have decided not to buy those services for everybody who needs them. Many people argue in favour of euthanasia (Section 23.7) on the basis of health rationing.

At the other end of the spectrum lies palliative care meaning 'specialised dying or medical treatment'. The main aims of palliative care service are: (a) to provide free consultation; (b) to provide medicines at subsidised rates to poor patients; and (c) to see that the patient need not suffer unnecessary pain. Doctors at Christian Medical College, Ludhiana, have launched Regional Palliative Services for the first time in the northern region.

23.7 Euthanasia

Euthanasia literally means 'good dying', which may sound like oxymoron. Many call it 'mercy killing' and some countries or states have also legalised it. However it is absolutely illegal in India as of now.

Pros of euthanasia

i. Euthanasia is a boon to the terribly suffering terminal individual.
ii. Given the serious constraint of healthcare delivery systems in India, this could be a suitable method for health rationing.

Cons of euthanasia

i. Who will decide which condition justifies euthanasia?
ii. What will be the safeguards and how will they be implemented to prevent abuses like personal rivalry or financial gains or eugenics?

Some suggestions

i. Doctors should serve and care for their patients compassionately, especially practice palliative care for the terminal patients.
ii. Deliberate attempt to end or shorten life, whether by omission or commission is unethical debatably, and should be restrained.
iii. The desperate wishes of the terminally ill must be duly honoured
iv. Moral and spiritual values must be instilled in healthcare personnel and people, which may lead to sound legislation.
v. Euthanasia may be legalised with proper safeguards against its malevolent use.

The end of life (non-vegetative existence) may be determined by brain death, which in turn may be a based on these criteria for switching off life-saving devices:

1. Unresponsiveness
   i. The patient is completely unresponsive to external visual, auditory, and tactile stimuli and is incapable of communication in any manner.
2. Absence of cerebral and brain stem function
   i. Pupillary responses are absent, and eye movements cannot be elicited by the vestibulo-ocular reflex or by irrigating the ears with cold water.
   ii. The corneal and gag reflex are absent, and there is no facial or tongue movement.
   iii. The limbs are flaccid, and there is no movement, although primitive withdrawal movements in response to local painful stimuli, mediated at a spinal cord level, can occur.
   iv. *Apnea Test:* An apnea test should be performed to ascertain that no respirations occur at a PCO_2 level of at least 60 mmHg. The oxygenation in the patient should be maintained by giving 100% oxygen through a cannula inserted into the endotracheal tube as the PCO_2 rises. The inability to develop respiration is consistent with medullary failure.
3. Nature of coma must be known
   i. Known structural disease or irreversible systemic metabolic cause that can explain the clinical picture.
4. Some causes must be ruled out
   i. Body temperature must be above 32 °C to rule out hypothermia
   ii. No chance of drug intoxication or neuromuscular blockade
   iii. Patient is not in shock
5. Persistence of brain dysfunction
   i. Six hours with a confirmatory isoelectric EEG (electroencephalography) or electrocerebral silence, performed according to the technical standards of the American Electro-encephalographic Society
   ii. Twelve hours without a confirmatory EEG
   iii. Twenty-four hours for anoxic brain injury without a confirmatory isoelectric EEG
6. Confirmatory tests (are not necessary to diagnose brain death)
   i. EEG with no physiological brain activity
   ii. No cerebral circulation present on angiographic examination (this is the principal legal sign in many European countries)
   iii. Brain stem-evoked responses with absent function in vital brain stem structures

23.8 Further Reading

1. www.consort-statement.org (CONSORT)
2. http://ohsr.od.nih.gov (Helsinki Declaration)
3. http://icmr.nic.in/ethical.pdf (Ethical Guidelines of ICMR)
4. Senn, S. J. Ethical considerations concerning treatment allocation in drug development trials. *Statist. Meth. Med. Res.* 11: 403-411. 2002.
5. Temple, R. and Ellenberg, S. S. Placebo-controlled trials and active-control trials in the evaluation of new treatments: Part 1, ethical and scientific issues. *Ann. Intern. Med.*, 133: 455-463. 2000.
6. www.ethics.ubc.ca
7. chrismac@ethicsweb.ca
8. http://fishwrap.mit.edu/Hacks/misc/ethics.html
9. http://ajobonline.com(Bioethics.net)
10. http://bubl.ac.uk/link

23.9 Questions

1. Descriptive Questions

1. How can a randomised controlled trial (RCT) be ethically made sound?

2. Write Short Notes on

1. Risk–Benefit Ratio
2. Healthcare Economics
3. Health rationing
4. Euthanasia
5. Ethical concerns for genetic manipulation

3. Multiple Choice Questions

Choose the single best response.

1. The specific issues of grave ethical concern include all except
   (A) Performance Engineering
   (B) Assisted Reproductive Technology (ART)
   (C) Stem-cell Research
   (D) Human cloning

2. An RCT with short-term versus long-term exercise as the intervention cannot be
   (A) open
   (B) single blind
   (C) double blind
   (D) triple blind

3. One of the means of achieving health rationing is
   (A) palliative care
   (B) euthanasia
   (C) both
   (D) none

4. Physical quality of life index (PQLI) does not include
   (A) GNP
   (B) Infant mortality rate (per 1000 live births)
   (C) Life expectancy at birth
   (D) Literacy rate (% of people above 15 years of age)

5. Present day ethical dilemmas include
   (A) Biodiversity preservation versus GMOs
   (B) Information confidentiality in EHR/PMRI
   (C) making guinea pigs of innocent uninformed people in poorer countries
   (D) all

Answers

1. A; 2. D; 3. B; 4. C; 5. D

Appendix 1

Glossary and Abbreviations

The Greeks had a word for it.

Zöe Akins

1. Prefixes

Intra	within (e.g., intracellular means within the cell)
Inter	between (e.g., intercellular means between the cells)
Hypo	less (than normal)
Hyper	more (than normal)

2. Planes of the body

Transverse: divides body into upper and lower segments

Longitudinal:

Frontal/coronal: divides body into anterior (front) and posterior (back) segments

Sagittal: divides body into right and left segments

3. Abbreviations

L	:	Angstrom = 10^{-8} cm or 10^{-10} m or 10 nm.
ADC	:	Analog (to) digital converter
ANN	:	Artificial neural network
ANOVA	:	Analysis of variance
ARMA	:	Aut regressive
BF	:	Blood flow
CES	:	Connectionist expert system
cf.	:	Greek meaning confer (compare)
CRBF	:	Cerebral regional blood flow
CRO	:	Cathode ray oscilloscope

CRT : Cathode ray tube (for display)
csf : cerbro-spinal fluid
CT or CAT : Computed tomography or computed axial tomography scan
CW : Continuous wave
DSA : Digital subtraction angiography
DTPA : See Gd
EBM : Evidence based medicine
ECG or EKG : Electrocardiogram or electrocardiography.
EEG : Electroencephalogram
eg : For example (*exempli gratia* in Greek)
EMG : Electromyogram
EOG : Electrooculogram
ES : Expert system
etc. : And other (things) (*et cetera* in Greek)
fMRI : Functional magnetic resonance imaging
FEM : Finite element method (of modeling)
FES : Functional electrical stimulation
FFT : Fast Fourier transform
GA : Genetic algorithm
Gd-DTPA : Gadolinium-diethylene triamine penta-acetate.
GTO : Golgi tendon organ.
H^+ : Hydrogen ion.
HA : Hydroxy apatite
HDD : Hard disk drive
HHS : Department of Health and Human Services (USA)
Hz : Hertz
i.e. : That is (*id est* in Greek)
ICA : Independent component analysis (new signal processing technique not discussed in this book).
IFS : Iterative function system
IV : Intra-venous
J : Joule
kPa : Kilo Pascal
LCD : Liquid crystalline display
m or mm : Micron or micrometer = 10^{-6} metre
MMA : Methyl methacrylate.
MRI : Magnetic resonance imaging
(N)MRS : (Nuclear) Magnetic resonance spectroscopy
OBG : Obstetrics and gynecology
OCR : Optical character recognition
ODE : Ordinary differential equation(s)
OH^- : Hydroxyl ion
PDE : Partial differential equation(s)
PEMA : Poly ethyl methacrylate
PET : Positron emission tomography
PMMA : Poly methyl methacrylate

PRF : Pulse repetition frequency
Q : Blood flow; (of a coil): ratio of inductive reactance to resistance
RBF : Regional blood flow
RF : Radio frequency
R/S : Receiver/Stimulator
SPECT : Single photon emission computed tomography
SNR : Signal (to) noise ratio
SSEP : Somato-sensory evoked potentials
T : Tesla
TOC : Third order cumulant
TFT : Thin film transistor
US : Ultrasound
USG : Ultrasonography
VDU : Video display unit
viz. : Vizier (like)

4. Definition of some important terms

Amplitude: The (vertical) magnitude of a signal measured as the height (zenith) of the signal from a zero reference level or baseline.

Bit: Binary digit

Byte: Eight (2^3) bits make a byte (usual word length). The higher or lower prefixes are in the powers of 2 instead of 10. e.g., kilobyte is not 10^3 (1000) but 2^{10} or 1024.

Capacitance: The ability to store electrical charge in a capacitor (or condenser). A condenser is just two conductors separated by a dielectric or insulator. The dielectric can be paper, mica, ceramic or plastic film/foil. High value capacitors are usually electrolytic. Electrical charge can be placed on the plates by connecting a voltage source to the capacitor. Capacitance is usually measured in microfarads (mF) or picofarads (pF).

Catheter: A hollow tube (usually flexible) of fibre, rubber or plastic and occasionally metal, introduced into a cavity (eg urinary bladder or heart chamber), vessel (eg PTCA or balloon catheter angioplasty in cardiac blood vessel), to allow injection or withdrawal of fluids for measuring the pressure.

Decibel: 1 dB = 0.001 W at 600W = 0.1 bel.

Diode: A vacuum tube or semiconductor having two electrodes, used mainly as a detector or a rectifier to convert AC (alternating current – variable instantaneous values) to pulsating DC (direct durrent – constant magnitude in one direction).

Feedback: Returning a part of the output, to the originating machine or system, for correction of error and control of the process.

FET: Field effect transistor or a voltage controlled semiconductor similar to a triode (three-electrode vacuum tube).

Filter: A voltage dividing network arranged to possess frequency discriminating properties.

Flip-flop: A storage device used to retain one bit of information.

Gain: This is the output to input ratio in an amplifying circuit.

Grid: An electrode comprising a cylindrical wire screen or spiral that is mounted between the cathode and vacuum tube plate to control the electron flow from cathode to anode.

Impedance: Obstruction to the flow of current. It is a function of frequency and a complex quantity.

Modulation: A variation in the carrier wave with time. The variation can be in amplitude (AM – amplitude modulation), with phase of successive peaks in phase modulation, in frequency (FM – frequency modulation) and in pulse modulation. The pulse of the modulated wave varies in step with amplitude fluctuations at a rate determined by its frequency.

Mosfet: Metal oxide semiconductor FET, comprising source and sink regions on either side of a P (+) or N (–) type channel along with a gate electrode insulated from the channel by silicon dioxide.

Noise: Undesirable signals (voltage or current interferences) within the signal/image.

Phase: The fraction of a cycle that any signal is away from a refernce point of the wave in question, at any instant. The phase difference between two signals is measured either in degrees or in fraction of a complete angular rotation.

Pulse: A momentary sharp change in current or voltage that may be recurrent but is not cyclic except in the case of a highly distorted waveform.

Semiconductor: A material having resistivity in between that of conductors and insulators. Over a specific temperature range, the electrical charge carrier comcentration is directly proportional to temperature.

Appendix 2

Reference Values

(old/ SI units with conversion factor or CF)

Blood pressure (reference adult male in rest): 120/80 mm Hg or 15.96/10.6 kPa (CF 7.52) A

Appendix 3

Greek Alphabets

α	alpha
β	beta
γ	gamma
δ	delta
ε	epsilon
ζ	zeta
η	eta
θ	theta
ι	iota
κ	kappa
λ	lambda
μ	mu
ν	nu
ξ	xi
o	omicron
π	pi
ρ	rho
σ	sigma
τ	tau
υ	upsilon
Φ	phi
χ	chi
ψ	psi
ω	omega

Appendix 4

Some General References and Important websites

Knowledge is of two kinds. We know a thing ourselves or we know where to find information about it.

Samuel Johnson

1. Leondes, C. T. Ed. *Handbook of Computational Methods in Biomaterials, Biotechnology & Biomedical Systems.* Kluwer Academic Publishers, Volumes 1–4. 2002.
2. Bronzino, J. D. Ed. *The Biomedical Engineering Handbook.* 2nd edition. CRC Press, Volumes 1–2. 2000.
3. Webster, J. G. Ed. *Medical Instrumentation, Application and Design.* 3rd edition. Wiley Student Edition, John Wiley & Sons (Asia) Pte. Ltd. 2004.
4. Myer, K. Ed. Standard *Handbook of Biomedical Engineering and Design.* New York: McGraw Hill. 2003.
5. Gerstner, W. and Kistler, W. *Spiking Neuron Models, Single Neurons, Populations, Plasticity.* Cambridge: Cambridge University Press. 2002.
6. Bentley, P.J. *Digital Biology.* Simon & Schuster. 2002
7. Potter, A and Leder, P. *Exploring the Biomedical Revolution.* HHMI, JHU Press. 2000.
8. Tozeren, A. and Byers, S.W. *New Biology for Engineers and Computer Scientists.* Prentice Hall Inc. 2000
9. Morton, D., Perry, J.W. and Perry, J. *Photo Atlas for Anatomy and Physiology.* Brooks/Cole. 1998
10. IEEE EMB Staff. Computers in Cardiology. IEEE Press. 1997
11. Bronzino, J. D. *The Introduction to Biomedical Engineering.* Academic Press. 1999.
12. Pacela, A. *Bioengineering Education Directory.* Brea: Quest Publishing Co. 1990.
13. Woodford, F. P. A Scientist's view of bioengineering. *Med. Eng. Phys.* 16: 91–96. 1994.
14. Schwann, H. P. Organizational development of biomedical engineering. *IEEE Eng. Med. Biol.* 10:

25–33. 1991.

15. *Dorland's Illustrated Medical Dictionary*, 28th ed. PA: W B Saunders. 1994.
16. Sarbadhikari, S. N. Let your dreams come true through virtual reality (Current Topic). *J. Indian Med. Assoc.* 95: 190. 1997.
17. Sarbadhikari, S. N. Health care delivery — The roads not taken (Editorial). *J. Indian Med. Assoc.* 93: 329–330. 1995.
18. Sarbadhikari, S. N. Biomedical engineering — Relevant or not? (Editorial). *J. Indian Med. Assoc.* 91(6): 145. 1993.
19. Sarbadhikari, S. N. Rehabilitation: The ultimate aim (Editorial). *J. Indian Med. Assoc.* 88: 184–185. 1990.
20. McCullough, NC III. *Atlas of Orthotics.* St. Louis: CV Mosby. 1975.

Websites for literature survey

1. http://www.ncbi.nlm.nih.gov/PubMed/(Medline Search)
2. http://www.medportal.com (Medline Search)
3. http://www.doctorspage.net/quotes.asp (Medical quotations)
4. http://cebm.jr2.ox.ac.uk/docs/searching.html (Centre for Evidence-Based Medicine, University of Oxford)
5. http://www.google.com (Search for 'Tutorial' on any topic)

Websites of international institutes

1. http://www.ieee.org (IEEE)
2. http://www.who.org (WHO)
3. http://www.hhmi.org (Howard Hughes Medical Institute)
4. http://www.ibro.org (International Brain Research Organization)

Websites for Indian biomedical organisations

1. http://sctimst.ker.nic.in/ (Sree Chitra Tirunal Institute for Medical Sciences & Technology (SCTIMST) at Thiruv-ananthapuram, Kerala, South India, is an institute of national importance under the Government of India)
2. http://www.nrdcindia.com/ National Research Development Corporation *(A Government of India Enterprise*)
3. http://www.medisourceasia.com/industrynews/
4. http://www.mcsindia.org
5. http://www.indmedica.com
6. http://www.icmr.nic.in (ICMR)

Websites for museums

1. http://www.si.edu (The Smithsonian Institute)
2. http://vh.org (Virtual Hospital)
3. http://members.evansville.net/ict/prostheticvalveimagegallery.htm

Websites related to biomedical engineering

1. http://www.tc.umn.edu/MDRCBB/ (Minnesota Dental Research Center for Biomaterials and Biomechanics)
2. http://www.bme.ri.ccf.org/ISB/ (International Society for Bio-mechanics, Cleveland Clinic Foundation)
3. http://www.hydroxyapatite.com/

1. http://www.devicelink.com/mpb/(Medical Plastics & Biomaterials – Magazine)
2. http://www.uchc.edu/biomaterials/home.html
3. http://www.surfaces.org/ (Symposium Workshop. Exploring surface characterization and problem solving techniques important in the manufacturing, development, and research of biomaterials)
4. http://www.swri.org/3pubs/ttoday/fall/implant.htm
5. http://www.utoronto.ca/IBBME/> (Institute of Biomaterials and Bio-medical Engineering)
6. http://www.biomaterials.ca/
7. http://odp.od.nih.gov/biomaterials/
8. http://www.biomed.abdn.ac.uk/Research/BioMaterials/ (Biomaterials & Bio-mechanics Research Group)
9. http://www.zoology.ubc.ca/labs/biomaterials/
10. http://www.ceac.aston.ac.uk/biomaterials/biomaterials.html
11. http://www.uml.edu/res/misc/intelbio.html (Center for Intelligent Biomaterials)
12. http://www.bioforma.com/sites.shtml
13. http://www.sciencedirect.com/science/journal/00219290
14. http://www.biomat.net/
15. http://www.biomaterials.org/
16. http://www.biomaterials.org.au/
17. http://www.ind.tno.nl/
18. http://www.ssb.biomaterials.ch/
19. http://www.whitaker.org/
20. http://www.bmenet.org/
21. http://www.ewh.ieee.org/soc/embs/
22. http://www.accenet.org/

Index

Symbols

A

B

C

D

O

P

Q

R

S

T

U

V

W

X

Y